LITERATURE AND THE PHILOSOPHY
OF INTENTION

Also by Patrick Swinden

AN INTRODUCTION TO SHAKESPEARE'S COMEDIES

PAUL SCOTT: Images of India

THE ENGLISH NOVEL OF HISTORY AND SOCIETY 1940–1980

UNOFFICIAL SELVES: Character in the Novel from Dickens to the Present Day

Literature and the Philosophy of Intention

Patrick Swinden
Senior Lecturer in English Literature
University of Manchester

First published in Great Britain 1999 by
MACMILLAN PRESS LTD
Houndmills, Basingstoke, Hampshire RG21 6XS and London
Companies and representatives throughout the world

A catalogue record for this book is available from the British Library.

ISBN 0–333–73499–8

 First published in the United States of America 1999 by
ST. MARTIN'S PRESS, INC.,
Scholarly and Reference Division,
175 Fifth Avenue, New York, N.Y. 10010

ISBN 0–312–21963–6

Library of Congress Cataloging-in-Publication Data
Swinden, Patrick.
Literature and the philosophy of intention / Patrick Swinden.
p.   cm.
Includes bibliographical references (p.   ) and index.
ISBN 0–312–21963–6 (cloth)
1. English literature—History and criticism—Theory, etc.
2. Literature—Philosophy.   3. Intention (Logic)   I. Title.
PR21.S95   1999
820.9—dc21   98-40373
                                                              CIP

© Patrick Swinden 1999

All rights reserved. No reproduction, copy or transmission of this publication may be made without written permission.

No paragraph of this publication may be reproduced, copied or transmitted save with written permission or in accordance with the provisions of the Copyright, Designs and Patents Act 1988, or under the terms of any licence permitting limited copying issued by the Copyright Licensing Agency, 90 Tottenham Court Road, London W1P 9HE.

Any person who does any unauthorised act in relation to this publication may be liable to criminal prosecution and civil claims for damages.

The author has asserted his right to be identified as the author of this work in accordance with the Copyright, Designs and Patents Act 1988.

This book is printed on paper suitable for recycling and made from fully managed and sustained forest sources.

10   9   8   7   6   5   4   3   2   1
08   07   06   05   04   03   02   01   00   99

Printed and bound in Great Britain by
Antony Rowe Ltd, Chippenham, Wiltshire

In memory of Logan Mitchinson
(1941–1998)

Contents

Preface ix

1 Good Intentions 1

2 Shakespeare 54

3 Coleridge and Kant 111

4 Milton, Sterne, Prince 172

Epilogue 230

Notes 235

Select Bibliography 250

Index 252

Preface

The aim of this book is to clarify some terms used by literary critics and theorists and to expose errors in their definition and application. Most of the terms belong to an area of discourse in which ideas about intention shade into ideas about motive and purpose, meaning to do things and saying what it is you mean to do.

It is also a study of the functions of purpose, motive and intention in literature. The operations of these psychological powers (if that is what they are) vary greatly, depending on whether we are attending to our own responses to the texts we are reading or to what we presume were the imaginative activities of their authors. Underlying the argument is the idea that the former is in some sense the counterpart of the latter. The creative power of the writer generates a corresponding reflective power in the reader, and this can best be studied as an imitation in the reader's mind of the intentional activity of the writer, as this is projected over the surface of his text.

The relationship between intentional activity and both a sense of purpose and a prompting motive is a large part of my subject. I believe that by examining this relationship it might be possible to drive a wedge between some powerfully influential, though malign, developments in contemporary literary theory, and readers' legitimate demands for an account of the activity of reading uncontaminated by notions of cultural determination and political relevance.

The role of authorial intention in the creation of great works of literature seems a good place to start. During the last twenty years or more, views of literary intention filtering down from modernist aesthetics have been scrutinized and found wanting. But the precise respects in which they have been found wanting have been obscured by developments that have tended to shift discussion of literary values onto a different ground. Where for the modernist the matter of intention was an obstacle to be dismantled, for the contemporary theorist it is merely an irrelevance, a sham. These contemporary developments might be conveniently described as deconstructionist, poststructuralist or simply postmodernist. They have in common a marked scepticism about the integrity of consciousness and a disbelief in the coherence of the human personality as a grounded, willing, responsible and intentionally alert subject in the world.

The challenge to humanist values that these views entail has become so extreme that it would be inadvisable for a literary critic to confront any of the main issues head on. But since the subordinate but related issue of intention has loomed so large in literary studies, it might be possible to conduct a flanking manoeuvre in that area. Some critics have already done this. But they have done it for the most part almost casually, as a by-product of doing something else. It is not surprising that this should have happened. A preoccupation with literary intention will often coincide with an inclination to focus attention on origins (study of sources and influences) and issues (analysis of the detail of the text). Only the anti-intentionalist will regard theory as a primary consideration. For him, both the origins and issues of what is unintended are viewed as evidence of much more significant, because extra-personal, activities that in some sense transcend them. Even so, recent years have seen the publication of several books placing contemporary developments in theory of mind and language in the service of an attack on post-structuralist assumptions about critical practice. Where appropriate, I refer to these in the conduct of my argument.

Another reason why the intentionalist case has gone by default is that the commonsense notion of intention underpinning the work of those who espoused it was incoherent. However, my argument is that the dismantling of current notions of how an author's intention makes texts work didn't entail a dismantling of the intentional character of the texts themselves. So critics were under no obligation to dismiss out of hand many of the notions routinely associated with the presence of authorial intention in the text. But postmodernist critics thought they were under such an obligation, because a spurious intentionality had been advanced as the chief support for ideas about responsibility, purposiveness, authorial identity, etc that had traditionally accompanied humanist criticism of literature.

The first thing that needs to be done is to look at the general definitions of intention that underpinned notions of more specifically literary intentions exposed as fraudulent by the anti-intentionalists. They have in common the idea that intention precedes action or that it accompanies action as a directive power; and that intention is in origin subjective and in expression either invisible or super-erogatory from the point of view of interpretation. Consequently intention is to be viewed not as a public but as a private characteristic of all human endeavour, and therefore of all human communication, including literary communication. In my opinion this is a mistaken view of the nature of intention. Contradicting it, I shall argue that intention is a non-subjective, public aspect of communication between human beings. It is so by virtue of the necessary coincidence of intention with all instances of the transactions it is possible to imagine one human being making with another or others. Verbal expression is one of those transactions. When it takes the form of a literary text, subtle and important changes are observable in the way the intentional force of what is expressed is related to the content of the expression. But intentional force itself is as evident and as necessary a feature of literary as it is of any other form of expression. In what follows I try to explain how and why this is the case, and how it has been felt to be the case by writers who are themselves often preoccupied with the intentional utterances of their characters as well as with the intentional character of the whole of the work of literature to which they belong.

The boundaries of the subject are difficult to establish. I have endeavoured to resist the temptation to transgress them, but have not been able to ignore altogether the larger implications of some of the claims I have sought to make. These have mainly to do with the degree to which a writer bears responsibility for the world of the text he creates, and with the relationships between powers of the mind which may in some sense be described as imaginative. I hope the argument advanced here goes some way towards establishing the importance of a writer's responsibility for his work and drawing attention to common features among forms of imaginative activity normally considered very different from one another. The combination of art and craft that attends both the creation and reception of literature, art and music is distinctive and unique, but not in the sense that these are the only activities in which the imagination plays a powerful part, or in which certain technical skills become inseparable from the way the imagination expresses itself.

Another boundary I have tried hard not to cross is the one between art and religion. It is not by chance that ghosts, witches, magicians and gods figure prominently in the following pages. All great art trespasses on the supernatural. Therefore the second-hand paraphernalia of humble criticism are bound to include a fair sample of the supernatural. But there is more to it than that. Supernatural figures in art are often the harbingers of precisely those ideas – of purpose and irrational meaning – which this book sets out to explore. The presence of these images of the irrational alongside a discursive argument about imaginative activity that is itself often considered (though in my view wrongly) irrational, might have the effect of diverting the argument into theological channels it was no part of my intention to navigate. That intention is entirely aesthetic in the sense of the word it is the purpose of the argument as a whole to clarify.

One crucial respect in which art is different from any other imaginative activity is that in it impracticality and purposiveness exist together and, in doing so, make possible the creation of beautiful objects. An entry in Wittgenstein's *Notebook* towards the end of 1916 suggests that 'there is certainly something in the conception that the end of art is the beautiful.' Kant thought so too. His definition of the beautiful, in the *Critique of Judgement*, places heavy emphasis on the sense of the purposive in art, while depressing the significance of precise intentions and definite motives. I have tried to make a similar point, and that is why the commentary on Kant's aesthetic critique that appears in Part III below lies close to the centre of my argument. My approach is by way of a consideration of some functions of intention and purpose in the plays of Shakespeare and in other genres of literature that occupies Parts I and IV of this book. Part I establishes a contemporary context for the argument. The Epilogue suggests, tentatively, through a metaphor abstracted from one of Ingmar Bergman's films, where that argument might lead. I have tried to clarify my procedures and definitions by references to contemporary philosophers of action, meaning and language.

An earlier version of the material on the poetry of Thom Gunn appeared in *The Critical Quarterly*, vol. 19, no. 3, 1977, pp. 43–61, and that on Milton in *Literature and Theology*, winter, 1998. I am grateful to friends and colleagues at Manchester with whom I have discussed the issues addressed in this book, even where many of them disagreed with the way I approached them. Especially my wife Serena, Betty Blanchet, Peter Clasen, Denis Crowley, Felicity Currie, Julie Edelson, Damian Grant, Richard Hogg, Peter Jackson, Grevel Lindop, Jonah Raskin, Alan Shelston and the late John Stachniewski. Much of the book was written during two sabbatical terms granted by the University of Manchester in 1992 and 1996. My thanks to Shelagh Aston, Sally Marshall, Maxine Powell and Mary Syner for the trouble they took with the typescript at various stages of its preparation.

Although all critics can reason more plausibly than cooks, yet the same fate awaits them.

> Kant, *The Critique of Judgement*

He who knows does not speak, he who speaks does not know.

> Lao Tse, quoted in Ogden and Richards, *The Meaning of Meaning*

1
Good Intentions

We lack words to say what it is to be without them.
P. F. Strawson, *The Bounds of Sense.*

THE PROBLEM OF MEANING

Strawson's epigram closes a rigorous analysis of Kant's account of the structure of experience. It refers to infants and non-human animals, making the point that we can only think of the experience of such creatures through a simplified analogy of our own.

> Any specific ascription of experience to animals we make involves thinking of them as perceiving this or that kind of thing, recognising this or that individual, pursuing this or that purpose in relation to such things. Any description we can give, any thought we can entertain, of *their* experience must be in terms of concepts derived from ours.[1]

It is a fact of life, though, that human beings are rarely satisfied with the analogies Strawson says they are bound to make. This has made them dissatisfied with words – or at any rate with words used in a fashion normally considered appropriate to their speakers as perceiving, recognizing and above all purposeful individuals.

One of the three distinctively adult human attributes Strawson refers to is a sense of purpose, and this does indeed seem to be a necessary part of any satisfactory definition of a human being. It also seems to be something that is necessarily excluded from the definition of any other sort of being. In the case of animals and babies it is used, as Strawson says, only by analogy, to make casual, non-scientific thinking about such creatures possible. Superhuman creatures, like gods and angels, are also habituated to our ways of thinking by having our ways of thinking attributed to them, and they too are conceived of as creatures having powers of deliberation, intention and purpose.

Pursuing a purpose, though, is different from having a concept of it. Some philosophers distinguish an intention from a purpose on the grounds that 'intention' is a more intellectual word and has closely associated with it the possibility of a declaration of intention. In *Thought and Action* Stuart Hampshire argues that 'intention' is, but that 'purpose' is not, out of place in descriptions of animal behaviour. He says:

> There are evident analogies between some kinds of animal behaviour and the purposive behaviour of human beings. A dog is seen scratching the ground, and to the question 'What is he doing?' a natural answer is 'He is playing at hunting a mouse.' Here it might seem as if the intention behind the activity is being stated, because at least the point and purpose of the activity are stated.[2]

The fact that Hampshire's dog is playing at catching a mouse doesn't mean that he knows he is catching a mouse. It is his master, not himself, who says that is what he is doing. The dog has no sense of purpose, let alone a concept of it. It is very difficult for human beings to separate having a purpose from having a sense of purpose. Indeed a sense of purpose comes close to being an intention, in Hampshire's 'intellectual' sense of the word.

All the same, experience tells us that though their meanings are very similar, having a purpose and having a sense of purpose are not quite the same thing. One would only speak of having a purpose when replying to a question about some particular activity. If I were a dog scratching the ground perhaps I would supply the answer Hampshire supplies about his dog, with the idea of a purpose in playing at hunting a mouse lurking somewhere behind it. The question 'What are you doing?' is often an invitation to state your purpose. 'What are you doing standing there?' 'I am waiting to catch the bus.' Having a sense of purpose, though, is usually a much weightier and more vaguely diffuse sort of thing, and one that animals almost certainly do not possess. Indeed, what has often been considered to differentiate human beings from animals is not that human beings pursue this or that purpose and animals do not, but that animals don't think about their pursuits of particular purposes as being related to any general purpose in life. Human beings conspicuously do have this sense of a general purpose, or purposefulness, in their lives. That is one of the reasons they often lack words to say what it is to be without them.

In modern times the ends to which a sense of purpose might be directed have grown more and more obscure. The thinning out, then the vanishing of goals derived from religious faith, political idealism or even liberal goodwill has left our sense of purpose uncomfortably exposed outside the parameters of its psychological origins and linguistic expression. Objective goals to which we might have felt our whole personality being drawn have changed their character, grown vague and ambiguous, and finally perhaps disappeared altogether. However, the separate purposes that, being animals, we continue to pursue still seem to want to fit into that more general sense of purpose that, being humans also, we are bound to experience. But whatever used to exist outside ourselves as a guarantee that the sense of purpose we experience has a value for most people exists no more. We have been left with a sense of purpose experienced as an efficient cause hanging in the void left where the same thing viewed as a final cause used to be. Consequently the feelings, hopes and fears associated with a sense of purpose have had to be projected onto a blank space. How, then, are we to use words – which, by virtue of their teleological character, their end-directedness, remind us of our identities as purposeful individuals?

The use of words is intimately bound up with the having of purposes and, therefore, the harbouring of intentions. So the attempt to divest oneself of the power of using words, or of using words somehow to get behind them to a state of being that has no need of them, is to try to divest oneself of the intentionality and purposiveness to which the use of words more than anything else testifies. Recent European literature affords many examples of writers trying to do just that: divesting themselves of the power of using words, in exchange for a wisdom grounded in silence.

The rationale of this ambition has not been at all times the same. Throughout the post-Romantic era the urge to move beyond language coexisted with a greater or lesser respect for ideas of purposive individuality and intentional action. At the beginning of the age, Kant faced in both directions. He was confident that it would be possible to produce a theory of mind and knowledge in which language would reflect the structure of the world as we understand it and, perhaps, as it is. But he also believed that poetry was first among the arts because the patterns created through the words that are its raw material carry the reader beyond the life of thought in which words are necessary or adequate. Poetry, he wrote, 'expands the mind by setting the imagination at liberty and by offering, within the limits of a given concept,... that which unites the presentment of this concept with a wealth of thought *to which no verbal expression is completely adequate* [my italics]' (*The Critique of Judgement*, I, 53, pp. 170–1).[3] Discussing 'Aesthetic Ideas' Kant often drives a wedge between language as a system of meaningful signs and human consciousness as a subject in control of that system. Nevertheless, this is not altogether characteristic of Kant's theory of art, and it is possible, even necessary, to offer an account of the theory which returns to poetry a proper concern with words as bearers of intentional meaning and purposive force.

Unlike most of his idealist successors, Kant was able to circumvent the negative implications of the view of human consciousness entailed by the passage quoted from *CJ*, and this had a salutary effect on his trust in the capacity of words to communicate meaningful statements. How different have been the attitudes to language of many contemporary poets and philosophers. Their repudiation of the pretensions of words has been more urgent and less circumspect. At the same time, they have been less concerned about the need to preserve ideas of intention and purpose in their writing. The poets especially have been hag-ridden with suspicion of words as bearers of clear and articulable meanings. Words have to be used to communicate truths that are finally beyond words, beyond language. Maeterlinck's King Arkel believed that the human soul was of its nature 'very silent'. Paul Valéry espoused the idea on several occasions in his *Cahiers*: 'La littérature essaye par des mots de créer l'état du manque des mots.'[4] Ted Hughes claimed he recognized 'a spirit, a truth under all the truths. Far beyond human words.'[5] This goes far beyond the fastidiousness one would expect writers and readers of poetry to feel about the slipperiness and ambiguity of verbal reference, and it has nothing to do with that mistrust of words evinced by poets who have sought unsuccessfully to cope with their experience of the barbarities of the first half of the present century. It has more to do with the connection between the use of words and harbouring of intentions, and, further, the presumed link between intentions and purposes that was of so much interest to Kant. 'All hastens to its end. If life and love / Seem slow it is their ends we are ignorant of,' wrote the American poet J. V. Cunningham. But few of his contemporaries have emulated his sanguine appreciation of the possibilities opened up by this combination of slow living and uncertain purposes. Instead, the situation seems to have struck them dumb, or, what is worse, forced them to explore, and then explain, the reasons for their dumbness. This is why their words are spoken so apologetically, as excuses for something else – something that can't be there, so that words have to be put there instead. The writing of a poem has often represented itself as an exercise in futility. It has been done to show it can't be done, and the doing of it becomes the undoing of it in the service of a higher, wordless cause.

THOM GUNN'S TOUCH

The poetry of Thom Gunn shows what happens to someone who has lost faith both in words and in the purposive and intentional character of human identity. In recent years his poems have grown fainter and fainter. They seem to be engaged in the nerveless exercise of rubbing themselves out. They look as if they are there only to tell their readers that they will soon be gone and that what appears on the page is nothing but the record of their going. One is left wondering why a gifted poet should have wanted to erase his poems in the very act of composing them. Above all, given the close connection between the use of words and the bearing of meanings – especially meanings of an intentional, purposive character – one is driven to speculate about the relationship between the kinds of poems Gunn has written and certain contemporary notions of identity, meaning and intention.

A poem in Gunn's first collection opens with the line 'I have reached a time when words no longer help'.[6] The problem, then, arose at the outset. From the beginning, his poems might be described as elegant queries about their right to exist, meditations on the inadequacy of language to express the related inadequacy of most descriptions of personal identity. Readers of these early poems, though, tended to think otherwise. At first the picture of the world Gunn was creating in *Fighting Terms* and *The Sense of Movement* seemed to be one in which words were used as instruments of self-definition. They were able to identify a robust, even aggressive, self-consciousness, standing free of whatever lay outside the circle it drew so tightly and securely around itself. This was not what happened in all of the poems. In 'Human Condition' individuality is both nameable, a 'pinpoint of consciousness' in the 'established border' of the poet's defining sensibility, and capable of metaphorical enlargement, a castle cut off from the world by its moat. But in the slightly later 'Vox Humana' the exercise is conducted differently. What establishes the speaker's identity in this poem is less certain and less easily articulated through clear images and precise definitions. It is not even clear whether the speaker is doing the defining, or someone else is defining him. He describes himself as 'an unkempt smudge, a blur/an indefinite haze, mere-/ly pricking the eyes, almost/nothing.' Identity is potential, a blurred intuition of what might be rather than a description of what certainly is. Though the image is insubstantial, it is also 'ominous'. It seems to have some sort of objective status as well as a subjective 'feel'.

Most readers have concentrated on Gunn's exercise of the existential choice consciousness has thrust upon him. But there is another side of him that wants to get back to the less clearly defined, less limiting and less strenuous state of being that was there before the duty of commitment to choice arose. This requirement also leads him to consider how he might get away from the poses, the descriptions, the separations of self from what is not self (his 'tangible presence' is felt to be 'illegal') in the teeth of the fact that these attributes seem to be written into the very words he uses to think in and communicate with. The springs of speech are different from speech itself. The 'right meanings' are found in the mouth of a lover who is 'silent'.

In his early poetry Gunn created a poetic landscape in which what was distinctively human was thrust painfully into a natural world of instinctive though hidden purpose, rank with growth and promiscuity. His people were most admired when they adapted themselves to the machinery of a man-made world set against nature. 'You control what you can, and/use what you cannot' is one of the confessions of the life artist in *Touch*. He looks back to the bikers of 'On the Move', 'us[ing] what they imperfectly control' as they 'scare a flight of birds across the field'. Action is urgent, wilful, clenched. But a significant change came over the poetry in the second part of Gunn's third volume, *My Sad Captains* (1961). Here Gunn scrutinized the detail of the natural world to find out how a relaxation of will and purpose might combine in a single structure both a formal integrity and a capacity for growth, change and dispersal. The syllabic poems in this volume are among the finest Gunn has written, but it is a condition of their being so that they fail to include several hitherto important aspects of his poetic character. They withdraw from personal contacts and relationships at the same time as they withdraw from the circle of self-consciousness the poet had spent so much time drawing and evading in the past. Their quiet visual concentration manages to assimilate what is seen to what is thought with no intrusion of more immediate and, perhaps, cruder sense impressions – of taste, smell and, above all, touch.

Where these senses had been solicited before, they too had been rapidly placed in the service of an idea, and so their crude physical appeal had been ignored. By releasing the intellectual brake on the solicitations of sense, Gunn achieved perfection within a very narrow compass. The release into what is other than self achieved in these epiphanic meditations was the other side of the poet's previous concern with various forms of self-examination, assertion and protection. However there was no way of permanently getting rid of 'the great obstruction of myself'. Self is not merely an intellectual presupposition upon which perceptions of other things depend. It is a palpable presence, locked into skin, flesh and bone. Experience of it is as much tactile as intellectual. Perhaps the best way to come to terms with this monster is through the medium of what is most grossly limiting and self-bounding – the body as a physical thing that occupies a certain space and position, and that is known to the man who moves in it most continuously and habitually by touch. The sense that most effectively turns the key in the lock on the other side of the door of the self might be used to unlock that same door from the inside. Gunn's next volume, *Touch*, is a record of his probings with that key.

The title poem shows the direction Gunn has turned in after the elegant dead end of the *My Sad Captains* syllabics. It appears in the first part of the volume and is written in a curiously unresonant sequence of five-syllabled lines with no rhyme. The first verse paragraph demonstrates clearly enough the unechoic, muttered quality of the poem:

> You are already
> asleep. I lower
> myself in next to
> you, my skin lightly
> numb with the restraint
> of habits, the patina of
> self, the black frost

> of outsideness, so that even
> unclothed it is
> a resilient chilly
> hardness, a superficially
> malleable, dead
> rubbery texture

After the first brief sentence, the paragraph drifts off into a leakage of qualification and almost inert personal description. It is easy to identify the 'patina of self' with the confining strictures of a regular verse form and a thrusting, purposive syntax. The self that is described in this way is 'numb with the restraint/of habits'. As the skin ceases to be numb, and the restraint of habits is lost, the sentences become less finished, less declarative. At the same time, the spurious rigidity of the poet's identity slackens. At first it seems the poem is saying, as poems usually do: '*I have imagined that* you are already asleep, and *that* I am lowering myself in . . .', with the result that these activities are thrust into the background of the poem, controlled and set at a distance by its shape. Really, the experience of the poem is already written before the actual writing has taken place. The rest of the poem seeks to bring that experience into the foreground, so that the present tense is felt to be really present, really a record of present sensing, questioning, being. As the resilience, the chill, the hardness of the body is invaded, warmed and softened, the movement of the poem becomes sluggish and intimately uncertain: 'Meanwhile and slowly/I feel a is it/my own warmth surfacing or/the ferment of your whole/body that. . . .' The words exist not so much to establish an identity as to lose one. The 'I' that is 'loosened' flows out of the singleness that seemed to be a part of its definition as a new being touched into life. It is 'an old/big place' already occupied by the poet's sleeping partner and his cat:

> What is more, the place is
> not found but seeps
> from our touch in
> continuous creation, dark
> enclosing cocoon around
> ourselves alone, dark
> wide realm where we
> walk with everyone.

There the experience ends, with the barriers of the isolated ego broken down and the self flowing out of it into a new, shared existence. Gunn has made his poem all foreground, all presence: what is going on in it is not an occasion (the sort of thing poems are made out of) but a habit (the sort of thing that is going on all the time).

In one of his fables Italo Calvino invented a language, 'Cimmerian', which is described as the last language of the living.[7] This is because books written in it are always unfinished, always continuing beyond, 'in the other language, in the silent language to which all the words we believe we read refer . . .' And because Cimmerian books *are* unfinished, Calvino's reader is vouchsafed only a perfunctory impromptu translation of one of their opening chapters. He never engages with the silent language beyond the Cimmerian words. But at the same time, he is tempted to demonstrate that

> the living also have a wordless language, with which books cannot be written but which can only be lived, second by second, which cannot be recorded or remembered. First comes this wordless language of living bodies . . . then the words books are written with, and attempts to translate that first language are vain.

There is a lesson here. What Gunn is trying to do in 'Touch' is to take Cimmerian English beyond its original terminus, to translate directly from the wordless to the worded language.

Other poems in *Touch* behave differently. They recognize the experience described in the title poem as an occasion, even an epiphany. In 'Back to Life' things are happening, not just discovering themselves to be as they have always been. The poet is walking aimlessly down a street skirting a park which is surrounded by lime trees and lit by lamps. At first he responds to the scene as an agglomeration of discrete particulars – a 'fragmentary shout', a 'distant bark' – and he sees himself as a 'patrolling keeper', alone and separate. Then he catches the smell of the limes 'coming and going faintly on the dark', and sees the leaves touching the lighted glass and rendered transparent at its touch. What appears to be merely sensory experience subtly alters the poet's feeling for the people around him:

> I walk between the kerb and bench
> Conscious at length
> Of sharing through each sense

> As if the light revealed us all
> Sustained in delicate difference
> Yet firmly growing from a single branch.

The next stanza makes it clear that this is only a temporary insight. Later, in rain and cold, it will be no more than a recollection 'held to by mere conviction' by a poet again locked in his single identity. Meanwhile:

> The lamp still shines.
> The pale leaves shift a bit,
> Now light, now shadowed, and their movement shared
> A second later by the bough,
> Even by the sap that runs through it:
> A small full trembling through it now
> As if each leaf were, so, better prepared
> For falling sooner or later separate.

The poem places the experience of coming back to life within a vividly conceived environment. At the end of it we can say: here we have a record of an experience that happened. This is very different from the end of 'Touch' where, I think, we must say: here we have the faint trace-marks left by a mind sinking into the sort of quiescence and unconsciousness out of which poems cannot be made.

In all of Gunn's collections of poetry since *My Sad Captains*, 'Touch'-style poems and 'Back-to-Life'-style poems have coexisted uneasily. The leakage of the first type congeals into something approximating to meaning and direction in the second, then floods out again in a formless dispersal of sensation. This has had a deleterious effect on both kinds of poem. The first has rarely entirely convinced as an apparently spontaneous set of responses to the self's indeterminate relation to the world. The fastidious command of syllabic verse forms tells against what it is there to express. The second has been undermined by the reader's suspicion that a man who scarcely exists at the centre of so much undifferentiating sensation cannot be expected to exercise the control required to bring that sensation within the bounds of poetic form. Instead, the shadow of his former *Fighting Terms* self takes over and produces a pale pastiche of what he had achieved in the days before self-dispersal had really got a grip on him.

HOLLOWAY ON STRUCTURALISM: DECONSTRUCTION

The belief in the power of the human will to make sense of the world in Gunn's early poetry is shared by many contemporary writers. Consider, for example, the structuralists. Their assumption too was that the mind and the world are in some fundamental way adapted to each other's requirements. In the early stages of their development, both Gunn and the structuralists assumed that language somehow reflected the structure of the world it describes. Since the aspiration was to demonstrate that this reflection is entirely accurate because the relationship is perfectly symmetrical, any flaw that appeared in the power of human will (in Gunn) and the fit between mind and the world (among the structuralists) might have been expected to have a proportionately devastating effect on the viability of the whole enterprise.

An English critic who has taken a sympathetic interest in structuralism traces these issues back to Kant. In *Narrative and Structure*, John Holloway refers to the 'Kantian terms' one might use to comment on the relationship between literary works and the world they describe. The forms and structures of experience 'in part represent the mind's own permanent nature and what in consequence are the standing possibilities of our consciousness', and these might in turn be reflected or repeated in literature:

> The *content* or the *detail* of literary works may often be seen to be what lies behind our being absorbed in the literary work... or what brings us transient moments of, say, intense poignancy or delighted surprise; but the experiences which seem to be the most massive and profound parts of our reading experience are other than these 'whodunit' excitements. They involve some total re-projection of the work: or some definitive falling-into-place of its overall perspective, which on analysis is likely to be the epiphenomenon of structure, or structure the epiphenomenon, maybe, of it.[8]

Holloway raises the question of the 'systemic relation between the general apprehension of reality by the individual consciousness, and its particular apprehension of the work of art'. What, he asks, is going on when a response is made to a work of art? To answer the question he introduces the idea of 're-animation'. A work of art, he says, is 'a creation of one individual consciousness re-animated, as it were, in apprehension by another individual consciousness which in fundamental ways is a twin of the first.' The Kantian assumption of an underlying identity of consciousness between one person and another is expressed through the idea of twinship. We can't help having knowledge, experience and sense-preferences that distinguish our own particular cast of mind from other people's, and since poetry and fiction appeal to personal experience and sensation, writers can't help drawing on their own possession of these things in composing their poems and novels. In so far as works of literature stake more general claims to our attention, though, they seek to penetrate beyond features of experience that encourage the selection and appreciation on the basis of personal preference. In Holloway's opinion, these features have nothing to do with the details of experience and sensation, or of the language that is used to communicate them, but with more formal attributes of the work as a whole that have been brought into being through a process of abstraction from the detail during the successive moments of the detail's composition.

Reading a poem or a novel, a person must impose restrictions on the propensity of his mind to go along with this process of abstraction. The link between the contents and details to which Holloway refers, and the text which comprises the sum total of those contents and those details, can never be entirely broken. Somehow the abstract formal structure and the detail of verbal nuance and represented subject have to stand together, even while their idiosyncratic and their universal references pull in opposite directions. As Kant says of empirical aesthetic judgements: 'A judgement of taste is therefore pure only so far as no merely empirical satisfaction is mingled with its determining ground. But this always happens if charm or emotion have any share in the judgement by which anything is to be described as beautiful.'[9]

The history of Gunn's poetry, no less than that of the 'corrections' of Kant's philosophy in the work of nineteenth-century Idealist philosophers, and the shift in Wittgenstein's thinking from the picture theory of language in the *Tractatus* to the word games of the *Investigations*, is symptomatic of a profound uncertainty in post-Romantic attitudes towards the 'fit' between word and concept and concept and world.[10] It is an exemplary instance of the way the idea of a stable and universally valid, though not necessarily isomorphic, correspondence between the structure of consciousness and the structure of the world (in Wittgenstein's case, between a proposition and the state of affairs it refers to or asserts) has been subjected to sceptical disparagement or, at the very least, severe modification. And this has had a deleterious effect on the idea of literary form as a reflection, or 're-creation' of that fact of correspondence. When the structure of correspondence broke down, Gunn's purchase on personal identity weakened, and his ability to write poems in the form of records of events of one kind or another faded away. He started to rub his poems out as he wrote them, much as Derrida actually used to cross out the words and phrases he used to articulate meaning in his writing about difference and deferral.

The problem these writers have addressed is: What is the identity and status of the 'I' we refer to, directly or indirectly, in the composition of a text? Who is the authorial subject, the one who appears to confront us directly in an autobiographical Romantic poem or diary entry, or who insinuates his presence with varying degrees of deliberation or evasiveness in a nineteenth-century bourgeois novel or even in a *nouveau roman* by Michel Butor or Alain Robbe-Grillet? Holloway asks the further question, a very interesting one, about which kind of 'fit' there is between the 'general apprehension of reality by the individual consciousness, and its particular apprehension of the work of art'. Structuralist critics of the 1950s and 1960s expected their linguistics-derived procedures would eventually discover what that 'fit' was. Conversely, the deconstructionists (who were often the same people at a later stage in the same inquiry) claimed that this would never be possible, because the terms of reference were inadequate. For them, there was and is no such thing as 'subject' or 'individual consciousness' to 'fit into' either reality or its representation in a work of art. Gayatry Spivak, introducing an early work of Derrida, writes of the need to unravel the notion of the author as the reader 'sees the text coming undone as a structure of concealment, revealing its self-transgression, its undecidability.'[11] It appears that there comes a stage in the reading of any text when the discovery of its self-contradictions exposes the artifice of the self that is contradicted.

Poststructuralist and Lacanian psychoanalytical literary theory share a rich vocabulary signifying different aspects and levels of subject-identity. Lacanians, for example, replace Freud's crude distinction between ego and id[12] with an assortment of linguistic signifiers – such as the subject, the ego, the *je* and the *moi* – which are often hard to differentiate from one another.[13] Together, though, they have the effect of shifting the emphasis from sexual impulses arising out of a distinct personal identity to what James Mellard calls 'functions of structured positions within a symbolic field'[14] – i.e. relations between units of an impersonal language system. The authority of the subject above or within the symbolic field is difficult to determine. Raymond Tallis claims, with some justice, that 'The Lacanian subject... seems unable to decide whether it is the shaping spirit... behind the imaginary or symbolic identifications; or whether, like the *I*, it is a product of those identifications. Whether, in short, it is *constituting* or, like the ego, merely *constituted*.'[15] Lacan makes plain throughout the *Ecrits* that human behaviour, including authorship, is a function of the unconscious, and that the unconscious is like a language and is to be described psychoanalytically in the same way as we describe the structure of a language. But this is a language which has severed its connection with 'the real world' of extra-linguistic objects and relations. For instance, the behaviour of the baby in Freud's *Fort! Da!* experiment is reinterpreted as having nothing to do with the real presence or absence of the spool he threw out of his cot, and everything to do with a principle of linguistic *differentia* represented by the contrasted sounds with which the baby has replaced it. The relations between the two sounds incorporate a meaning that no longer requires the memory or the concept of an object to substantiate it. And the continued throwing and rewinding of the spool in and out of the cot has itself become a symbolic act having more in common with the operations of language than extra-linguistic involvement in the real world.[16]

Derridean deconstruction also directs the reader's attention to operations of language which appear to refer to objects and relations outside the text but which in fact refer back to the disruptive power of the text itself. Derrida's celebrated insistence that there is no *hors-texte* runs parallel with Lacan's view that the subject is to be identified with the unconscious, and that the unconscious is structured like a language. But in both cases the sense of unity, inadmissibly brought into existence by false conceptions of how language works, fragments under the pressure of the appropriate 'reading' skills displayed in Derrida's grammatological practices and Lacan's linguistic psychotherapy. It is important to emphasize the double direction in which the phrase *hors-texte* points: both outwards, to the world the text appears to represent, and inwards, to the unified consciousness through which the representation of that world is deemed to have been constructed. Both of these representations are spurious, because both substitute an extra-linguistic and discrete spectacle or instrument of perception whose existence is not warranted by the facts.

For Derrida, the effectiveness of our reading will depend on whether we approach the text as a book or as writing. Michael Payne explains the difference in relatively simple, intelligible terms:

> To consider a text... as a book presupposes a unified structure and authorial control of meaning that becomes clear as one reads. Whatever tensions or conflicts are generated by the structure of the book, these are ultimately resolved once the overall design and intention are fully perceived. Unity is the basic structural principle of the book. To consider the text *as writing*, however, liberates it from presuppositions of static, objectified, authorial structure and recovers the sense of signifying and reading as complementary processes.[17]

Perhaps this is too simple, and too readily intelligible. What it is saying is that traditional definitions of reading invest heavily, even exclusively, in notions of 'authority' that the writer exercises over his material and that he induces the reader to share indirectly in the process of reading the text. Deconstruction, on the contrary, encourages a more equal and cooperative approach. At first, Payne's interpretation of the deconstructive alternative seems unexceptionable and rather tame, pitting stuffily rebarbative phrases like 'static, objectified... structure' against more energetic and liberating ones like 'complementary processes'. But there is much more at stake than this. For the deconstructionist, authority over the text is shared by writer and reader to a degree that collapses the distinction between production and consumption. The text ceases to be the invention of one person that is in some way reinvented through the interpretative activity of another, and becomes instead a network of unlimited interpretative possibilities. The possibilities are unlimited because interpretation is a function of aspects of language and 'subject'-definition that, according to theorists like Lacan and Derrida, are extraordinarily fluid – characterized as they are by the complementary facts of division of the psyche and arbitrariness of the signifying system. So what at first look like reconcilable characteristics of the text as book and the text as writing are really not reconcilable at all. The differences are too extreme, because they derive from fundamentally opposed ideas about the relationship between mind, language and the external world.

In this book I intend to treat deconstructionist approaches to the text as literary historical and theoretical facts.[18] My only other concern with deconstruction is to draw attention to the way it privileges aspects of texts that are bound to escape the notice of readers going about their business as conscientious interpreters, i.e. rightly or wrongly behaving as empirical selves bringing experience to bear on their response to and interpretation of what lies before them. This might or might not include notions of transcendental selfhood, to be secured or deconstructed as the occasion arises. But it certainly excludes the acknowledgement of a sort of prior self-division to be matched against the bifurcations of the signifying process and what Payne calls the 'plurivocity' of the text. For the point at which Derridean deconstruction and Lacanian psychoanalysis join is the claim that the subject is discontinuous, consciousness is structured like a language, and language is non-referential. This triple claim, fundamental to all deconstructive theory, produces the description of a self that is simultaneously incoherent and aware of its incoherence through a propensity to frame mocking and irrelevant comparisons with 'comforting myths of wholeness' (as Ragland-Sullivan calls them)[19] – myths like the univocal sentence of pre-Saussurean linguistics, or memories of the false Imaginary reflection of Lacan's mirror stage.

At the heart of the deconstructive enterprise is the notion that consciousness is fundamentally linguistic. Deconstructive critics oppose the Kantian view that language is just one among several explanatory structures we work within, or tools we use, or media we exploit, when we apply ourselves to solving particular problems or asking general questions about the world. They oppose it on the grounds that the last word on the subject of language/meaning relations was implied, if not articulated, in Saussure's *Cours de linguistique générale*, and Saussure's view was that though the linguistic sign unites sound-image and concept, nothing in the sign derives its meaning from, or necessarily refers to, facts about or objects in the world.[20] The reason they (not Saussure) believed this was that they failed to distinguish between the value and signification of words, and between the sense and reference of propositions. The result of their believing it was that consciousness of anything beyond signs and sign systems became at first provisional, then problematic, then entirely spurious. Consciousness approximated to one of two things: either the mysterious phenomenological entity to which Lacanian adults occasionally regress when they are drawn to the Imaginary reflection of themselves caught at the mirror stage, or the adult symbolic order in which everything is internal to language. The first is an illusion, the second a form of linguistic captivity. There is no place in either of them for the 'design and intention' excluded from Payne's description of the text *as writing*. This represents what we must suppose is the final stage in a fading sense of self-identity that Thom Gunn arrived at from a different direction at about the same time as Lacan, Barthes and Derrida were making their first impressions on the Anglo-Saxon world.[21]

The place where words no longer help is only occasionally a figure of transcendent awareness of superhuman reality. When Eliot failed to speak to the ambivalent figure of the hyacinth girl near the opening of *The Waste Land*, the silence at the heart of light he looked into reflected the fact the he 'knew nothing'. In 'Burnt Norton', the 'overheard music hidden in the shrubbery' was associated with an empty alley, a drowned pool, later a pool that was empty. Yet there also the surface of the pool 'glittered out of heart of light', an image as deceptive as that earlier one of silence, dumb reflections and illusory brilliance. A later poet, Geoffrey Hill, claimed that in Eliot's poetry 'the expansive, outwards gesture towards the condition of music is a helpless gesture of surrender, oddly analogous to... that desire for the ultimate integrity of silence, to which so much eloquence has been so frequently and indefatigably devoted.'[22] Silence, wordlessness, seems to gesture towards an incommunicable reality. But in the end, if reality is incommunicable, it must fall outside the scope of translatable human experience and thus be without meaning. Meaning is always meaning for *us*; it is always transitive. That is who the unheard melodies of Keats's urn are ultimately unconsoling, the inexpressive gift of his 'cold Pastoral' to the breathing human passion it transcends. It is why the 'music heard so deeply that it is not heard at all' of 'The Dry Salvages' has to be incorporated, mystically, into the gift of Incarnation, before it becomes comprehensible as anything more than one of the 'hints and guesses' the poet fails to understand at the end of this third *Quartet*. At the opening of 'Burnt Norton', Eliot had expressed this incomprehension through the image of words detached from intuitions of time (especially) and space. The words echoed in the mind of an audience scarcely distinguishable from the poet himself. He in turn failed to discover an identifiable and intentionally active self within the spuriously differentiated times and tenses of the verse. Echoes are disembodied words, insignificant words, reverberating beyond the meaningful confines of an individual voice, person, being:

> But to what purpose.
> Disturbing the dust on a bowl of rose-leaves
> I do not know.

A sense of personal identity can crumble under the pressure of an unresolvable conflict between the exertion of will and the distractions of sense, or as a result of the subjection of mind to the impersonal authority of codes and symbols over which it has no control, or because of intuitions of experiences that transcend such habitual categories of thinking as those of agency, intention and purpose. It may be that on the other side of this breakdown of significant speech there is a land of lost content, a world full of whatever it is words fail to encompass – pure sensation, supra-personal freedom, real presence, ineffable religious joy. But all the evidence we have is words, and of necessity they cannot express the inexpressible. Gunn's 'Back to Life' has a human content that is lacking in 'Touch', and this seems to me to account in large part for its superiority as a poem. Eliot's poetry is most effective where it insists on the fitfulness, incompleteness and inadequacy of its mystical illuminations. Indeed, it seems to be a condition of their appearance in his verse that these illuminations *are* fitful, incomplete and inadequate. The *scriptible* texts of the *nouveaux romanciers*, contrary to familiar report, are not at their best when surrendering a claim to represent anything beyond the sheer act of writing. They are most impressive when they discover new formal arrangements to express feelings like jealousy and sexual violence (in Robbe-Grillet's early novels) or, less obsessively, 'our gradual, remorseless progress towards death' (in those of Claude Simon).[23]

Poststructuralist critics would deplore this substantiation of Eliot's and Robbe-Grillet's practice. For them it is an illusion. The writer has no personality beyond the grammar of the words that reflect the fact of his cultural determination. All he is is words, and the concepts the words conventionally stand for, and the relationships between concepts that the rules governing the associations of words make permissible. 'Linguistically, the author is never more than the instance writing, just as *I* is nothing other than the instance saying

I: language knows a "subject" not a "person" ...'[24] That is what Barthes showed us Balzac was in *S/Z*. And that is all any writer can be, unless he is also someone who knows this is the case, in which event he builds that knowledge into the structure of his writing – like Robbe-Grillet tracing the soldier's search for his parcel in *Dans le labyrinthe*, or failing to resolve his heroine's uncertainties about what, if anything, happened *L'Année dernière à Marienbad*. Abrams refers to Derrida's 'all-or-nothing principle'.[25] Lacking a ground in absolute presence or the transcendental signifier, and thus failing to meet absolute criteria which language cannot satisfy without ceasing to do its work, 'all spoken and written utterances, though they may give the effect of determinate significance, are deconstructable into semantic indeterminacy'.

The failure of language to achieve intentional meaning is predicated on the absence of a transcendental subject to validate its claims. Sean Burke has drawn attention to this 'rigid and rigorous division... between the transcendental presence of intention and its no less transcendental absence' in Derrida's theory.[26] There appears to be no halfway house between presence and absence of intention and subjectivity absolutely and interdependently conceived. Burke congratulates post-Nietzschean phenomenologists like Freud, Heidegger, even Foucault, for promoting 'new forms of subjectivity through the refusal of the kind of individuality which has been imposed on us for several centuries' (Foucault's words), discovering 'a de-alienated subjectivity no longer split between transcendental and empirical essences, between a sovereign *cogito* on the one hand and an impenetrable unthought on the other.'[27] But the way these phenomenologists have sought to heal the split has resulted in very little useful commentary on the specifically literary uses of language. Perhaps the insistence on the word 'essence', and all the metaphysical baggage that accompanies it, creates the problem. Both phenomenological and poststructuralist thinkers are by definition anti-essentialists; but their philosophies are hag-ridden with nostalgia for the essences they seek to expose and eliminate. Tallis points to the extraordinary tautology of Husserl's conception of 'presence' as 'something that is given to, or returns to, itself without being "mediated"; an absolute proximity of acts to themselves'. He concludes that 'It is scarcely surprising... that there is nothing in ordinary experience answering to it.... The failure of the phenomenological quest to find absolute presence does not imply that presence in the ordinary sense is a concept riven by contradiction or that presence itself is an illusion and a myth.'[28] If that is true of Husserl, how much more true is it of those more recent phenomenologist heroes who appear in Burke's list, and how much more does it imply about the longing for unmediated presence that is so ineffectually disguised in their efforts to undermine the false notions we are supposed to entertain about it?

The instabilities of meaning, misdirections of reference and ambiguities of interpretation that are necessary properties of poststructuralist writing are merely outward manifestations of the fundamental incoherence of a world existing nowhere outside the text. Unmediated presence is a logocentric illusion. Signs are never grounded in a non-signifying beyond-sign. None is a radiant self-presence like Qfwfq's sign in Calvino's story, which was 'the thing you could think about and also the sign of the thing thought, namely, itself.'[29] The intra- or extra-subjective reality that structuralists like Todorov believe are signs are signs *of* lies nowhere outside the system of signs themselves.

To make some sort of sense of the fact that many readers do, nevertheless, derive satisfaction and pleasure from 'readerly' texts, we need to preserve the structuralists' faith in a stable relation between consciousness and formal expression, and between formal expression and the structure of the world. One way of doing this is to get rid of what Tallis calls 'isomorphic mapping' or 'physical synecdoche' as a grounding for such faith. More positively, we need to substitute something for whatever we identify as the falsifying principle structuralists share in the way they go about defining the relation. My argument is that this falsifying principle is the supposition that it is possible to separate the idea of man as a signifying from the idea of man as an intentional creature.

In an argument about *Beginnings* in literature, Edward Said comes to a similar conclusion. For him, 'beginning is a consciously intentional productive activity', and it is misunderstood by the structuralists because in their work 'the moving force in life and behaviour, the *forma informans*, intention, has been...domesticated by system'.[30] Stuart Hampshire writes that 'The often quoted fact that human beings are essentially thinking, and therefore symbol-using, animals is a special case of the fact that they are essentially intentional animals.'[31] Of course, being 'essentially intentional' doesn't entail being unremittingly so. In an essay on 'The Wit and Wisdom of Clarendon', Christopher Ricks points out that 'Those who still hold to the belief that it is possible for a reader to take an unaccountable pleasure in a writer's words are not obliged to equate intention with fully conscious decision: felicities are a coinciding of the happily unplanned with the deliberated.'[32] We can afford to take a relatively relaxed view of the connection between intentional and decisive literary activity, so long as we are scrupulous in discriminating between them. But only by binding together signification and significance in a conception of man as a fundamentally intending, purposive *and* signifying creature are we likely to be able to preserve the 'structuralist' faith in the relation between self, text and world. Only when we take account of the intentional character of the signs we use will the combination of those signs come to signify more than the reflections they throw back to us of our own incoherence, conventionality and irresponsibility. How, then, should we go about replacing these signs as members of a self-bounded and reference-free system into engines of intentional activity and purposeful direction in the world at large?

THE INTENTIONAL FALLACY: NUTTALL'S REMARKS

The American 'New Critics' advanced different arguments for separating the writer from the text. We need to examine their views in some detail because they placed even more emphasis than the French on the importance of notions of intention. Indeed one of the seminal essays of the New Critical school, in Wimsatt and Beardsley's *The Verbal Icon*, is, famously, called 'The intentional fallacy'.[33]

Wimsatt and Beardsley's central statement of their position is that 'the design or intention of the author is neither available nor desirable as a standard for judging the success of a work of literary art, and it seems to us that this is a principle that goes deep into some differences in the history of critical attitudes'.[34] They make three claims. One, the intention of the author is not available, there is no way of finding out what it is. Two, even if there were, it would have no bearing on the value judgement they would make on a work of literature which had in some sense issued from it. Three, the history of modern criticism largely revolves around the question of authorial intention and its relation to objective judgements about the value of works of literature. This third claim is unexceptionable. But I believe the other two need more support than Wimsatt and Beardsley provide, and that the deconstructionist critic Paul de

Man was right when he claimed that 'the partial failure of American formalism... is due to its lack of awareness of the intentional structure of literary form.'[35]

To counter them, it is necessary to disengage their use of the word 'intention' from suppositions about how the meaning of that word is to be universally understood.[36] There are several of these. The first is that intention is always private. If it were not, it would be available. Wimsatt and Beardsley make a firm distinction between the internal and external evidence for the meaning of a poem. Internal evidence − of linguistically and culturally determined features of a poem (such as its semantics and syntax), and our intuitive knowledge of the language in which it is written − is public. External evidence − 'revelations... about how or why the poet wrote the poem' − is private. They acknowledge there is a third 'intermediate' kind of evidence about an author's character, where this affects the 'private or semi-private meaning he attaches to the words or topics of his poems'. Clearly this is an area requiring scrupulous attention from the critic. If he does his job properly it will be obvious to the reader whether the information available within this third area belongs with the public internal evidence of the poem or the private external evidence of the poet's psychology.

The weakness in the distinction lies elsewhere, in the assumption that the 'how' and the 'why' of writing are coincidental. I shall argue below that a 'how'-question and a 'why'-question belong to very different categories of inquiry into intentional behaviour. While it is true that, as Wimsatt and Beardsley argue here, the answer to a 'why'-question will in all probability be provided through reference to external evidence, it is not self-evident that the same is true of a 'how'-question. Wimsatt and Beardsley have conflated two different meanings of 'intention' in their descriptions of external evidence, and this seriously weakens their argument about what they take to be the single, monolithic meaning of that word.

Wimsatt and Beardsley take Livingston Lowes' approach to Coleridge as their example of the critical misuse of author psychology. However deep he were to dig into the literary and philosophical sources of 'Kubla Khan', they argue, a critic would not be able to identify the origins of the poetic impulse that produced it. Ignore for the moment the proposition that, even if he had been able to do this, Livingston Lowes would have provided no help at all in evaluating the poem. The first point Wimsatt and Beardsley are making is that authorial intention is by definition undiscoverable, and therefore interpretation of the poem, as well as evaluation of it, must be conducted along quite different lines from those that look for some kind of consonance between intention and achievement. They come to this conclusion for the same reasons as those according to which, earlier, they identified intention with an event or events in the poet's inner consciousness. Intentions are absolutely private. They are what Nigel Williams' Inspector Rush said they were when he expressed the view that 'what counts... is our intention. What counts is our secret thoughts.'[37] Eric Griffiths refers to this view of the matter as one which 'conceives intention as an interior state of the subject's consciousness', operating 'behind' utterances rather than 'between' them. Intentions are misinterpreted as 'occulted inner experiences' rather than 'peceptible design'.[38] For Wimsatt and Beardsley, when they are successfully realized in a poem, intentions lose their character as intentions and become just part of the poem: 'If the poet succeeded in doing what he tried to do, then the poem itself shows what he was trying to do'. The question of intention simply doesn't arise. On the other hand if he fails, 'then the poem is not adequate evidence, and the critic must go outside the poem – for evidence of intention that did not become effective in the poem'. Wimsatt and Beardsley make it clear that in doing this the critic stops being a critic and becomes instead a Romantic biographer. So intentions that are realized in poems and intentions that are not realized in poems are equally irrelevant to a proper reading of poems as poems.

Concealed in this description of poetic success is the assumption that the poem on the page is an object, entirely cut off from the verbal activity that produced it. 'Judging a poem', Wimsatt and Beardsley say, 'is like judging a pudding or a machine. One demands that it work. It is only because an artefact works that we infer the intention of the artificer.' A poem, then, is an object, not an act; a completed thing, not a release of verbal energy that is always in process of *becoming* a thing in the reader's mind as he reads it. Drawing attention to this misconception, Allan Rodway tries to clear it by substituting the word 'purpose' for 'intention', defining 'purpose' as 'the apparent intention of a work as it reveals itself to the reader from the title onwards.' One of the principal advantages of this substitution, he claims, is that 'the concept of Purpose counters the tendency set up by many critical terms.... to take the work as an *object*, whereas it is more importantly an *action*.'[39] Concealed in the description of poetic failure is the assumption that the line between failure and success is boldly drawn; mysterious and untraceable biographical intentions on one side of the line, complete transformations of such intentions into a poetry that has severed all connection with intentionality on the other. But the realization of a complex intention, or perhaps of a series of simple but related intentions, is rarely if ever complete, total and unequivocal. There is something suspect about the way Wimsatt and Beardsley separate off from each other what the poet tried to do and what the poem shows he was trying to do, only to stick them back together again retroactively in the description of the poem that issues from the act of trying.

Everyone agrees that it would be misguided to seek a one-to-one correspondence between operative and programmatic intentions, between something the poet has written down and something he had in his head immediately before he did it. No one supposes there can be a direct, unmediated transference of thought and feeling from one person to another. Wittgenstein pointed out that 'it does start to get quite absurd if you say that an artist wants the feelings he had when writing to be expressed by someone else who reads his work.' But he added 'Presumably I can think I understand a poem (for example), understand it as an author would wish me to – but *what he* may have felt in writing it doesn't concern *me* at all.'[40] T. S. Eliot looks as if he agrees with Wittgenstein on several occasions in his essays – most persuasively, perhaps, when reflecting on his own poetry. Writing about 'Prufrock', he insists that by saying a critic's analysis interested him, he is *not* saying that 'he saw the poem through my eyes, or that his account has anything to do with the experiences that led up to my writing it.'[41] Yet it could be argued that there is a world of difference between the 'feelings' Wittgenstein and the 'experiences' Eliot refer to, not least in respect of the extent to which it is suggested that either of them exists coincidentally with or anterior to the act of writing. It is doubtful whether Eliot would have conceded it is proper to speak of feeling in descriptions of the act of writing, and it is equally doubtful whether Wittgenstein would have accepted Eliot's implication that experiences that lead up to writing and the experience of writing are one and the same kind of experience. However, neither feelings nor experiences are species of intention. They just have a propensity to get involved with descriptions of intention when critics are considering poetry, because poetry obviously has something to do with feelings and experiences, and people like Wimsatt and Beardsley appear to go about with a picture in their heads of intentions, feelings and experiences all in some way or other preceding the act of writing and then being excluded from the accomplished written act. But Wittgenstein also wrote '"I am not ashamed of what I did then, but of the intention which I had." And didn't the intention reside also in what I did?'[42] So here the intention doesn't precede action, as a cause precedes an effect. It accompanies the action. In a sense it is inside the action – 'in what I did'. More recently, Robert Audi has described intentions as 'dispositional states' which 'do not cause anything except by virtue of some event suitably connected with them, such as their becoming occurrent', which, presumably, might happen during the writing of a poem. Donald Gustafson, elaborating a theory of intention hingeing on the concept of psychological causation, has suggested that 'an agent's doing some thing intentionally does not require that antecedently he "is (was) intending" that thing.'[43] All of this is very different from the causally active and temporally anterior intention Wimsatt and Beardsley were writing about.

'The Intentional Fallacy' was first published in 1946, and much water has flowed under the bridge since then. The concept of intention has been carefully examined in the philosophy of mind and the philosophy of action, and though there is no more settled positive opinion in this than there is in any other field of philosophy, it would be true to say that there is a settled negative opinion to the effect that it is a mistake to describe intentions as causes, and either it is a mistake to describe intentions as actions or it is necessary to be more than usually circumspect in so doing. Until recently, however, these developments in philosophy have had little impact on literary criticism, even in the work of critics like E. D. Hirsch, P. D. Juhl and Stanley Fish,[44] whose practice is based on considerations of authorial and other types of intention. They don't seem to have had much effect, either, on the 'New Pragmatic' argument of Steven Knapp and Walter Ben Michaels in their seminal essay 'Against Theory', which has dominated discussion of authorial intention since the mid-1980s.[45]

Knapp and Michaels argued that the meaning of a text is identical to the author's intended meaning and that therefore the project of grounding meaning in intention is incoherent.

> The mistake made by theorists has been to imagine the possibility or desirability of moving from one term (the author's intended meaning) to a second term (the text's meaning), when actually the

two terms are the same. One can neither succeed nor fail in deriving one term from the other, since to have one is already to have both.

Meaning is just another word for expressed intention, and neither phrase is able to explain or add anything to the other. Both intentionalist and anti-intentionalist critics have taken up very defensive attitudes towards this argument. Intentionalists are fobbed off with authorial intentions they can do nothing with, and anti-intentionalists are deprived of the satisfaction of using non-intentionality as a means of deconstructing textual meaning. On the other hand, many anti-theoretical critics have been happy to use the New Pragmatic argument as a sort of clearing operation preliminary to getting on with business as usual.

All three parties are mistaken, because as William C. Dowling points out in his answer to Knapp and Michaels,[46] the argument is based on a logical error. It is not true to say that meaning and intention are identical. Instead, we should say that meaning and intention entail each other. Knapp and Michaels confuse identity and mutual entailment. They fail to identify 'a situation in which either of a pair of terms is rendered meaningless when treated in isolation from another.' In Dowling's opinion, this isn't a serious objection to Knapp and Michaels, because identity and mutual entailment behave so much alike that the central New Pragmatic argument survives their blurring of the distinction. At the level of theory, this might be true. But in practice, I shall argue, the substitution of entailment for identity makes a lot of difference, because it leaves the way open for a point of entry into the text which is something other than a direct approach to its meaning. Even if considerations of intention necessarily entail inquiries about meaning, they take a different form from direct questions about meaning, and this allows the person who is doing the considering to reflect on aspects of meaning that might escape a full frontal approach. Dowling suggests as much when he adds:

> Since meaning must always in purely formal terms involve intention (as opposed to intention as the prior psychological state of a deceased author) and since where there is an intention there must always be an intender, interpretation must begin by positing something like the internal voice and speaker of formalist theory.

As Knapp and Michaels point out in their reply to Dowling, nothing in his account of the internal speaker rules out an external authorial intention. This is true. And if the external authorial intention cooperates with the internal speaker (or speakers) as it does in the work of Shakespeare, Milton, Sterne and Prince examined below, it will become clear that identification of such intention is essential in enabling the reader to grasp its meaning.

A recent essay on the subject appears in A. D. Nuttall's *The Stoic in Love*. It is called 'Did Meursault mean to kill the Arab? The intentional fallacy fallacy'. Nuttall reproduces the question Wimsatt and Beardsley ask about 'How is [the critic] to find out what the poet tried to do?' and their reply that 'If the poet succeeded in doing it, then the poem itself shows what the poet was trying to do.' It is extraordinary how seriously critics take this self-evidently absurd riposte, which depends for its effectiveness on the false equation of trying to succeed and succeeding in having tried. As the analogy with the pudding suggests, the particular kind of trying and succeeding that Wimsatt and Beardsley had in mind seems to presuppose a situation in which first one thing seemed to happen (the poet had something in his mind that he wanted to try to do) and then another thing happened (he did it). Failures don't count. The whole or the part of the poem that succeeds is then the effective result of the causal operation of placing whatever was in the poet's private mind into the public arena, thus making it over into an object for the scrutiny of others.

This is what Nuttall calls the dualist view of intention, according to which 'in every intentional performance there is, first, a pre-existent, identifiable relevant state of mind and, second, an overt action having some appropriate relation to the state of mind.' It follows that 'success in fulfilling an intention can be measured only by the degree of correspondence between the state of mind and the public action. No-one who was ignorant of the prior state of mind could measure this correspondence.'[47] For obvious reasons, this remains the preferred definition of intention in the philosophy of law. 'Intention is the purpose or design with which an act is done. It is the foreknowledge of the act coupled with the desire of it.... An act is intentional if, and in so far as, it exists in idea before it exists in fact.'[48] Nuttall applies this dualist view of intention to two situations, one of which he has made up, the other being the one referred to in the title of his essay.

The one he has made up is a little story about a man with a gun shooting a terrorist who is threatening his wife. As Nuttall points out, it is possible that the man shot the terrorist after thinking to himself the repeated proposition 'If he gets close I'll shoot him', and this will satisfy the dualist argument for intention as a precedent cause of an action that subsequently occurs. But it is the essence of the dualist case that this should be the only, necessary and sufficient argument for intention. What, then, if the man doesn't think the appropriate and repeated proposition to himself, and instead experiences 'simply a series of disconnected images and reflections "The light's gone bad . . ."' etc? [Nuttall's list shadows the details of the story he has invented.] Does the shooting of the terrorist become something other than an intentional action? Not at all, Nuttall argues: 'In most of the action quite properly called intentional there need be no corresponding scheme in mind at all.'

We discover this, too, in reading Camus' *L'Etranger*. Meursault is not exonerated from acting intentionally in killing the Arab just because Camus chooses not to insert an explicit statement of intentional action between a description such as: 'Meursault had armed himself in case something went wrong, and when something did go wrong he fired five times'; and another one such as: 'During the incident his head was full of light and heat of the sun, of feelings of oppression, flickering anxiety and memory, awareness of a knife drawn before him, desperation.' This is Nuttall's own way of articulating alternative descriptions of the event. The first one is what he describes as a quasi-behaviourist description of a public action; the second a description of a state of mind that doesn't include the sort of intentional proposition represented in the 'If he gets close to me I'll shoot him' type of sentence. Either one of these descriptions points to the intentional character of the act Meursault performs. Either one of them will do, on its own, as an explanation of what happened that includes the intention of the agent. They can be, but they don't *have* to be, used together. Above all, the second one doesn't have to be used as a causal explanation of the first, and it wouldn't have had to be even if it had been replaced by the explicit intentional proposition 'If he gets close I'll shoot him', or some such statement.

In terms of this 'quasi-behaviourist' approach to intention, is there any difference between shooting an Arab and writing a novel about it? At first sight there seems to be a lot in common between Nuttall's explanation of what meaning to say or do, or intending to say or do, means. He gets rid of a superfluous term in the description of intentional action, and in doing so makes it possible to accord a status to intention as a significant presence in the actions he describes. But are we satisfied that in both cases the way we, the spectators, apprehend the intention in the action is the same? In Nuttall's second example, do we register Meursault's intention in killing the Arab in just the same way as we register Camus' intention in describing Meursault killing the Arab?

Nuttall has expelled the notion of agency as something existing separately from what an agent does and how an agent behaves. He has got rid of one kind of manifestation of what Gilbert Ryle called 'the ghost in the machine'. Nuttall claims to have shown that 'we do not need to be admitted to the introspective field of the agent before we decide whether his action is intentional or not; one always has [decided] and always will decide such questions by referring to public criteria.' This is a very Ryle-like view of intention, as Nuttall admits (though Ryle is oddly reticent about intention in *The Concept of Mind* – the word doesn't even appear in the index), and Ryle is one of the most influential of the philosophers referred to above as being of potentially great assistance to the literary critic, even when they do not themselves practise, or seek to lay secure foundations for, literary criticism. To take account of what Ryle and others have written about intention is a great help in the task of expelling many of those delusions that have beset literary critics both before and after Wimsatt and Beardsley came on the scene. But it has not dispelled them all. Indeed in some respects it has added to the confusion. Part of the unsatisfactoriness as well as the good sense of Nuttall's reply to Wimsatt and Beardsley can be attributed to the use he makes of post-Ryle philosophies of mind and action. We shall need to look more closely at some of these philosophies to see how they might help us to refine the argument about concepts of intention and motive when applied to a field of endeavour to which their authors scarcely ever refer.

RYLE, AUSTIN, GRICE, STRAWSON

The basic aim of Ryle's philosophy of mind is to clarify our thinking about mental concepts by reducing their incidence in our descriptions of human actions. Most of the words we use to describe actions, and the concepts they refer to when we use them properly, are appropriate and necessary. But it is more often the case than not that we use too many of them. This is mainly because the 'picture' of the mind that we imagine to ourselves, and the sketch of its operations we produce, mislead us into supposing that at any given moment at which an action is being performed, two things are going on one after the other instead of one thing going on that is capable of being regarded under more than one description. The two things that are supposed to be going on are usually conceived of as a cause and an effect. The cause is entirely mental, and the effect is a mentally directed action. Of course the mental cause might itself be pictured as if it were a physical action, because the way we have become used to thinking about mental events during the past four hundred years owes more to the study of mechanics than to any other branch of knowledge. The picture of the mind most people in Ryle's day and our own are likely to recognize resembles the one accompanying the opening titles of a BBC psychiatric soap opera: a little man is performing emotionally and intellectually expressive movements inside the outline of his own head. The image is similar to one that Ogden and Richards supplied as an illustration of traditional ideas of the soul: 'Before we looked carefully into other people's heads it was commonly believed that there is a little man inside the skull who looks out at the eyes, the windows of the soul...',[49] or to Lacan's 'little man in the big man' representing the ego in popular versions of Freudian theory.[50] There is a double mental operation: one belonging to me, and one belonging to the me who is inside my head. Theoretically this is an infinitely regressive interpretation of the process of thinking, because there is really no reason why I shouldn't be wondering what other sort of little man is performing what other sorts of activities inside the little man's head. But the theoretical problem is clear enough with just one little man in one big man's head. Is this really how we are best advised to picture the way our minds work when we are performing intelligent mental operations like deliberating, planning, judging, wondering and intending?

Ryle claimed that this is what we tend to do, and that we have been encouraged to do it by philosophers who have been unable to see though the fraudulent epistemological premises of Cartesian rationalism and the British empiricists. His opening chapter is called 'Descartes' Myth' in order to emphasize the antiquity of the problem and to point to one of the most celebrated but misguided attempts to produce a solution to it. Descartes' *cogito* precedes and ratifies his *sum*, presenting to us an early display of the ghost in the machine, which elsewhere in Ryle goes under a variety of spectral disguises: 'ghostly harbingers', 'occult episodes', 'shadow performances' and 'tandem operations'. I call them spectral because the metaphorical epithets are often derived from the world of the supernatural, where it is impossible to imagine a phantasmal presence without some reference to a non-phantasmal one that is in some sense its counterpart. One way or another a single thing has been split into two things, and the two things are conceived of as separate from each other. A ghost is not just another way of describing a dead man, whose ghost he is, come to life; the man, or at least the body of the man, is somewhere else, and he has caused the ghost to appear. The significant difference in Ryle's way of looking at it is that for him the direction the metaphor faces is reversed, and the immaterial ghost has become the cause of mentally directed activities of both mind and body.

We should rid ourselves of these importunate and redundant spectres. For in Ryle's view, 'when we describe people as exercising qualities of mind, we are not referring to occult episodes of which their overt acts and utterances are effects; we are referring to those overt acts and utterances themselves.'[51] We tend to think we are referring to 'occult episodes' because when we see someone performing an action and look for the point or meaning of the action – i.e. judge in what respect it is an intelligent action in the sense of being directed to some end or goal – we are bound to look beyond the performance itself. How otherwise would we differentiate between the action of an intelligent adult and that of an 'idiot, a sleepwalker... or even, sometimes, a parrot', or any other performance executed accidentally or mechanically? But to speak of 'looking beyond the performance' is also, inevitably, to speak metaphorically, though the metaphor is more effectively disguised in the verb than the spectral metaphors were disguised in the epithets – and for that reason it is all the more likely to mislead. Ryle develops the metaphor in theatrical terms:

> In looking beyond the performance itself, we are not trying to pry into some hidden counterpart performance enacted on the supposed secret stage of the agent's inner life. We are considering his abilities and propensities of which this performance was an actualisation. Our inquiry is not into causes... but into capacities, skills, habits, liabilities and bents. (45)

It is as if Hamlet's soliloquies were to be construed as rehearsals of subsequent performances instead of dramatic representations of a state of mind accompanying and affecting the performances themselves. (This does not exclude the presence in them of the activity of articulating plans for performances to follow.) In fact there is no counterpart activity or secret stage. There is only the act itself, the performance itself, viewed under descriptions that include more than the mechanical and accidental details of its operation. This 'is not going beyond in the sense of making inferences to occult causes; it is going beyond in the sense of considering...the powers and propensities of which [a person's] actions are exercises' (49).

How does this apply to intentions? Ryle doesn't explain in so many words. In his chapter on 'Emotion' he offers three designations of the word, the first of which is 'inclinations' or 'motives'. (The others are 'moods' or 'agitations' or 'commotions'.) These are differentiated from 'feelings' by virtue of the fact that they are not occurrences and therefore don't take place either publicly or privately. 'They are propensities, not acts or states' (81). Ryle gives us no reason to identify motives and intentions with each other, and I don't propose to do that either.[52] It may be, though, that the best way to understand what intentions are for someone who shares Ryle's general approach to the description of mental events is by way of his definition of motive.

Motives do not cause actions. They are what it is, in the character of an agent, that accounts for his having acted in a particular way on a particular occasion. 'The imputation of a motive for a particular action is not causal inference to an unwitnessed event [in the mind of the agent] but the subsumption of an episode proposition under a law-like proposition' (87). The use of the phrase 'subsumption under' is important. The action doesn't follow after the motive. It is the event or episode whose character is modified by the perception of a motive that is deemed to have accompanied it. It is an adverbial modification of a verb, translated from grammatical into conceptual terms:

> The sense in which a person is thinking what he is doing, when his action is to be classed not as automatic but as done from a motive, is that he is acting more or less carefully, critically, consistently and purposefully, adverbs which do not signify the prior or concomitant occurrence of extra operations of resolving, planning or cogitating, but only that the action taken is itself not

absent-mindedly but in a certain positive frame of mind. [Then] The description of this frame of mind need not mention any episodes other than this act itself, though it is not exhausted in that mention. (107)

The choice of the word 'purposefully' here, as one of the few motive-adverbs, is significant. It is one of those words – like 'deliberately' and 'directedly' – that are commonly applied as much to intentions as to motives. For example, it is in this sense that the ubiquitous distinction between an arm moving and moving an arm is explained with recourse to descriptions from mechanics on the one hand and [human] actions on the other. Alasdair MacIntyre does this in a paper, 'The Antecedents of Action', that brings into close proximity what Ryle says about motives and what MacIntyre believes might usefully be said about intentions.

There may be cases where we first frame an intention, come to a decision or experience a desire and then act; but the concepts of intention, decision and desire are equally applicable where the action is itself the expression of intention, decision or desire and to refer to our intention, decision or desire in either explanation or justification of our action is not to refer to an antecedent event.[53]

Like Ryle's motive, MacIntyre's intention 'unlike a cause, does not stand in an external, contingent relation to an action' (215). And intentions then take their place with purposes, decisions and desires as among those indications of a 'positive frame of mind' (Ryle) that serve to make human actions intelligible. But they don't function as causes of a human action, because they are not events, and they are not therefore antecedent to the actions which they modify by giving them a directed, intelligible character.

On the basis of MacIntyre's analysis it would appear that in a world governed by Ryle-like explanations of actions, intentions and motives have a great deal in common. Of course, there are distinctions to be drawn. Intention and motive are not identical.[54] But they are the same in so far as they are often identified as variants of the ghost in the machine and as such they are often held to account in some way for the activities that are associated with them. In point of fact they are more properly to be identified with the activities themselves, viewed in a certain light, under certain kinds of description – where those activities are being judged as intelligibly directed towards a goal of human consciousness.

Apart from one or two references to Jane Austen, Ryle doesn't say much about reading books. (One gains, rather, the impression of a man who spends his time watching cricket, buying provisions at the local grocer's and, of course, contributing to philosophical discussions.) But there is an interesting account in chapter 2 of *The Concept of Mind* of what it is to be a spectator (of a play) or reader of a book. Needless to say, such a person doesn't make

> analogical inferences from internal processes of his own to corresponding processes in the author of the actions or writings...He is merely thinking what the author is doing along the same lines as those on which the author is thinking what he is doing, save that the spectator is finding what the author is inventing.

In metaphorical terms, 'the author is leading and the spectator is following, but their path is the same' (54).

Ryle anticipates the objections of the anti-intentionalists when he mocks 'the pretensions of historians to interpret the actions and words of historic personages as expressions of their actual thoughts, feelings and intentions' (a rare use of the word in Ryle). The pretensions of the historians are in this respect identical with the pretensions of scholars and critics, as Ryle himself points out. But the problem disappears if the artificial gap between 'expressions of their actual...intentions' and the intentions themselves is closed up, the expression of their thoughts being delivered with an intentional force directed at a specific goal or goals. For 'if such penetration is impossible, the labours of all scholars, critics and historians must be vain; they may describe the signals, but they can never begin to interpret them as effects in the eternally sealed signal boxes' (55). If the signal boxes are not sealed, however, and the levers and the points they are supposed to control are in fact one and the same thing, differently viewed under variable aspects, then it becomes perfectly possible to interpret them on the lines of Ryle's leading and following down the pathway of the text. And it also follows that Eliot's and Wittgenstein's worry about interpretations of the text in terms of the communication of private experiences or feelings can be set on one side. There is no sense in which 'by re-enacting your overt actions I re-live your private experiences' (56), even where the overt actions take the form of written texts. (Ryle is talking about

Plato at this point.) The intentional activity taking place at the same time as, and taking the the form of, whatever one does when one writes out of one's private experience, makes the activity as a whole, and the text which is in a sense a record of the activity, a public one.

What Ryle doesn't examine here are the implications of the present indicative form he uses when he says that 'The author is leading and the spectator is following' when their activities are the same, viewed under a different description. He implies that the respect in which writing or reading a text is directed towards a goal is the same as that in which being responsible for any other distinctively human activity is directed towards a goal. Maybe most useful ways of reading Plato are like that, but some are not – and this accounts for the irritating distinction that is often made between reading Plato as a philosopher and reading him as a literary artist. Certainly, reading Ryle's favourite Jane Austen doesn't strike me as being much like reading most philosophers, as a goal-directed activity. And in this instance it is the philosophers who are performing the commoner action of the two. There is something about reading Jane Austen or Shakespeare that marks it off from most other human activities, when they are not themselves influenced by aesthetic considerations of whatever is being done. Under certain descriptions, of course, they might very well be influenced by such considerations.

From Ryle we have derived the proposition that intentions, like motives, accompany actions and add particular kinds of force and power to the directedness of actions. Two such actions are writing and reading a text. They have a lot in common with each other. It goes without saying that reading is subsequent to and dependent on writing. So in order to discover how intentions add force to the articulation of a written text, we need to know how the act of writing the text is to be defined in the appropriate conceptual terms. Here Ryle cannot help us. Instead, we shall have to turn to philosophers of language, and apply their definitions of verbal or linguistic activity within the wider parameters of Ryle's theory. We shall have to keep firmly in mind Ryle's view that 'the style and procedures of people's activities are the way their minds work and are not merely imperfect reflections of the postulated secret processes which were supposed to be the workings of minds' (57), and then apply it, with whatever modifications are deemed necessary, to descriptions of utterance-activities in general and literary utterance-activities in particular. 'Style and procedures' sound as if they will be appropriate terms to use here, so long as they continue to be understood as referring to the way minds work and not to aspects of the representation of the workings of minds.

Any use of language by a human agent is an illocutionary act. The phrase is J. L. Austin's, and I shall define it by referring to John Searle's essay 'What is a Speech Act?'[55] Searle provides a definition of a speech act which takes account of the earlier formulations of both Austin and Paul Grice. It is the most satisfactory working definition we are likely to find, not least by virtue of the fact that it is sufficiently general to make it possible for us to set to one side certain difficulties arising when some of the terms used are more minutely analysed and reformulated:

> In the performance of an illocutionary act the speaker intends to produce a certain effect by means of getting the hearer to recognise his intention to produce that effect, and furthermore, if he is using his words literally, he intends this recognition to be achieved in virtue of the fact that the rules for using the expression he utters associate the expression with the production of that effect.[56]

To begin with, Austin had not used the phrase 'illocutionary act'. Instead, he had drawn a distinction between 'constative' and 'performative' utterances. Constative utterances were the making of statements. Even at this early stage it was evident to Austin that in the past philosophers had attributed far too great an importance to constative utterances as being the type of an utterance of which all other utterances in their different ways fell short. This was unsurprising because constative utterances seemed to include all the utterances philosophers themselves used in the construction of their theories about the world. Nevertheless, they were a very small proportion of the sum of all possible utterances. Most utterances are performative. They are responsible for articulating the activities of promising, advising, warning, betting, ordering, *etc* which are activities that in the normal course of our lives we spend much more time performing than we spend in making statements and formulating propositions.

By the time he published his William James lectures as *How to Do Things with Words*[57] in 1962, Austin had realized that in dividing utterances into constative and performative he had fallen into the trap of simplifying the character of the language people use. It now transpires that all utterances are acts. Constative utterances do things just as performative ones do. So all speech acts, as it is now permissible to call all utterances, are performatives. Consequently Austin's later philosophy, and that of subsequent philosophers of language, took the form of an examination, or a tabulation, of different categories of performative utterances. Searle's emphasis on the performance of an illocutionary act indicates something of the extent of the agreement that has been reached, because all of the most influential philosophers of language start from the assumption that their business has to do with explicating the meaning and uses of this term.

Towards the end of *How to Do Things with Words* (p. 109) Austin distinguishes 'a group of things we do in saying something', as a preliminary to compiling a list of explicit performative verbs. The first member of the group is a locutionary act. This, he says, is roughly equivalent to 'meaning' in the traditional sense, i.e. 'uttering a certain sentence with a certain sense and reference'. The second is the illocutionary act Searle mentioned in his definition of a speech act. This is the performance of utterances like informing, ordering, warning, undertaking (Austin's examples) 'which have a certain (conventional) force'. Notice that these utterances include functions that would have fallen under the definition of a constative, as well as of a performative verb, in the old terminology. These first and second members of the group are present in all utterances, in all speech acts. The third, which is the perlocutionary, act may or may not be present. This is not, like an illocutionary act, what we seek to bring about or achieve *in* saying something, but what we do bring about or achieve *by* saying something. The examples Austin provides are convincing, persuading, deterring, surprising and misleading. Having made this triple distinction among different activities that may be or must be included in all speech acts – in his own words, the 'different sense or dimensions of the "use of a sentence"' – he makes two interesting qualifications. The first is that 'of course' there are other ones that he has not sought to explain. The second is that all three kinds of actions are 'subject to the usual troubles and reservations about attempt as distinct from achievement, being intentional as distinct from being unintentional, and the like'. Both of these qualifications will play an important part in our application of speech-act theory to the reading of imaginative literature. But the reference in the second to reservations about intention suggest that, in the light of Searle's own use of the word (three times) in his definition of illocutionary acts, more needs to be said about this matter, especially in the light of Austin's own distinction – eagerly taken up by later practitioners of this kind of philosophy – between intentional and conventional aspects of illocutionary acts.

Austin talks of illocutionary acts as having 'certain (conventional) force'. They are essentially conventional. By this he means that an illocutionary act such as arguing or warning can be made explicit in a sentence where the formula 'I argue that' or 'I warn you that' precedes the presentation of the argument or the warning. He contrasts this with a perlocutionary act such as convincing or alarming, both of which imply something about the actual as distinct from the intended effect of the illocutionary act. We can't say 'I convince you that' or 'I alarm you that' in the first person present indicative, because that would be to substitute illegitimately an effect for an intention. I think it is true to say that this use of the word 'conventional' in order to distinguish between illocutionary and perlocutionary sense of a speech act is generally agreed among philosophers of language. But Austin wants to go further than this. He suggests that all illocutionary acts are essentially conventional by virtue of the fact that the force of an utterance is determined by the nature of the circumstances in which it is performed.

There is a simple and there is a complex interpretation of Austin's meaning here. The simple interpretation is applicable to illocutionary acts which are performed in particular circumstances that account for the special force that attends them: for example, the words spoken by the bride and groom at the wedding ceremony, or the umpire's giving the batsman 'Out!' This is easy to understand. But there is also a more complex interpretation which has to do with a more subtle and farther reaching definition of what constitutes a conventional circumstance. Here there is no external circumstance to carry the burden of conventional signification. Instead, the conventional nature of the utterance resides in its subject's lack of what a post-Austinian philosopher has called a 'complex intention' 'secur[ing] understanding of the illocutionary force of his utterance'.[58]

Austin thought illocutionary acts are bound up with effects in three ways: by securing uptake, taking effect and inviting a response. Each of these is an aspect of the intention to produce some sort of response in an audience, or a readership, and that intention is sufficiently well served by the speech act's conformity with what Austin calls 'conventions of illocutionary force'. These are conventions which do not require an audience's recognition that, though the illocutionary has had an effect on them, this is the effect the author of the act intended it to have. Grice and Strawson and others have countered this essentially conventional account of speech acts with one or other alternative account that restores to them their fully intentional character. On the foundation of Grice's analysis of meaning 'as intending to produce an effect in a hearer by getting him to recognise the intention to produce that effect',[59] Strawson writes of the illocutionary act as going beyond 'simply an intention to produce a certain response in an audience', to 'an intention to produce that response by means of a recognition on the part of the audience of the intention to produce that response, this recognition to serve as part of the reason the audience has for its response, and the intention that this recognition should occur being itself intended to be recognised'.[60] By this means the person responsible for the speech act will have 'secured understanding of the illocutionary force of his utterance' and 'performed the act of communication he set out to perform'. He will have done so by drawing his audience's attention not only to the locutionary sense and reference of what he has said, but also to the fact that he said it with the conventional force of an intention, and also by including in the act the more restricted, and transparent, intention to secure an appropriate response. This last is the 'complex intention' that differentiates Grice's and Strawson's account of intentional or conventional illocutionary acts from Austin's account of essentially conventional ones. In each case the intention referred to is a feature or aspect or part of the act itself.[61] There is no Wimsatt and Beardsley intentional ghost in the machine that produces a speech act. How, then, can we apply the concepts and terminology of Austin and his critics to those special types of speech acts that go into the making of literary texts and poetic fictions?

We have seen that Austin doesn't exclude the possibility of there being more senses or dimensions in the use of a sentence than are indicated in his threefold scheme of locutionary, illocutionary and perlocutionary speech acts. On the contrary, he takes it for granted that there are more of them. But it is difficult to see what they might be. Certainly there is no need to assume that in moving away from the 'host' speech acts of ordinary language towards the 'parasitical' speech acts of the language of literature (the metaphor is Strawson's) we shall need to import a terminology that augments the one

Austin invented in *How to Do Things with Words*. What we might have to do, though, is place a different sort of interpretation on the relations obtaining between the several aspects of the speech act that Austin has already identified. In particular, we might feel inclined to depress the significance of the locutionary and perlocutionary aspects, and coincidentally emphasize that of the illocutionary aspect.[62] And we might do this most successfully and most appropriately by locking into our descriptions of the illocutionary force of a speech act both the intentional and the conventional features that Austin, Grice and Strawson were, in their different ways, so eager to separate.

In their demonstration of illocutionary force, the speech acts of imaginative literature behave quite differently from those used in the non-narrative transactions of everyday life or those of science and philosophy.[63] This is a consequence of the quality of disinterestedness which has been identified as a uniquely characteristic feature of the arts from at least as far back as the British aesthetic philosophers of the eighteenth century. Imaginative literature uses language metaphorically in the sense that what the words mean is not the same thing as what they ostensibly refer to, and it uses language impractically in the sense that the primary effect the works are calculated to have on the reader is not that of changing his actions, behaviour or even attitudes. Therefore its claims to locutionary and perlocutionary status are somewhat etiolated. Of course such claims are advanced, by the author, the reader, the critic – even, it might be argued, by the text itself. But there is a certain legerdemain in the way this is done. What makes us read literature has much more to do with its illocutionary force than with its locutionary sense and reference or its perlocutionary effect on our feelings, thoughts and actions.

I suspect that, with certain reservations, most readers would agree with the first of these propositions. It is to the second that they are most likely to take exception. If literature doesn't have an effect on our feelings and thoughts (perhaps not actions, they will agree, remembering Matthew Arnold), what does it do? The answer, I would argue, in the terms laid down by Austin and the speech-act philosophers, is that it acts on our thoughts and feelings without affecting them in the sense of producing a significant or lasting change in their character. It is true that in so far as literary texts, in common with all other texts, comprise locutionary acts, this will have an effect on the reader's thoughts. If it does nothing else, it will produce more material for them to work on. And the illocutionary characteristics that I am arguing play so significant a part in literary speech acts will often have the same sort of force we are accustomed to responding to in non-literary speech acts: they will seek to argue, to warn, to command, and this may in turn have the perlocutionary effect of convincing, alarming and causing to obey.[64] But what is characteristic of literary texts that is *not* characteristic of other texts or any other speech acts is an illocutionary force which has a manifest priority over Austin's other two dimensions of the speech act. Clearly the illocutionary force of literary speech acts cannot exist separately from those other dimensions. Nevertheless the way a speech act secures uptake and invites a response differentiates its operation in imaginative literature from its operation in all other forms of written and spoken communication.

Both the conventional and intentional aspects of an illocutionary act play a part in the way its force is registered by somebody reading a poem or a novel. By 'conventional' aspect I refer uncontroversially to both of the senses of the phrase that Austin defines and Strawson accepts, setting aside Austin's more comprehensive claims. The first of Austin's requirements of an illocutionary act – among those that define its character as being conventional – was that it should be capable of being made explicit in a performative formula: 'I agree that' or 'I warn you that', for example. The fact that what is performed is an act of arguing or an act of warning is displayed in the illocutionary force of the speech act or the sentence. In a sentence used in the course of day-to-day living, or in a textbook, the illocutionary act will add some sort of colouring, some kind of modal emphasis, to the locution. And that in turn might both secure uptake and succeed in having a perlocutionary effect on the audience beyond the fact of the audience's merely registering what is said.

The difference between the modus operandi of the illocutionary force here and in the numerous speech acts that constitute a poem or a work of fiction is that in the second category the locution is made to serve the interests of the illocutionary act rather than vice versa. Since there will be very many speech acts in even a simple lyric poem, this means that the sense and reference of the poem act as a sort of control or restraint upon the free play of its illocutionary tendencies. Perhaps this is not far off what Coleridge meant by 'the balance and reconciliation of opposite and discordant qualities', 'modifying a series of thoughts by some one predominant thought or feeling',[65] or I. A. Richards by the gloss on Coleridge's theory of the Imagination in chapter 32 of *The Principles of Literary Criticism*. In any event, the writing of the poem is conducted in defiance of ordinary expectations about what the relationship between locutionary and illocutionary aspects of a speech act is and should be.

Another sense in which Austin and Strawson agree that illocutionary acts are conventional is where the circumstances in which the speech act is performed display 'established conventions of procedure additional to the conventions governing the meanings of our utterances' (Strawson, p. 26). Searle is referring to a similar conventional use in the type of speech act he describes as a 'declaration'. It is the type of speech act that includes the words spoken at the marriage ceremony – 'I do' – or by the foreman of the jury – 'Guilty, m'lud'. These words spoken on these occasions have a special force that conforms to conventions that are built into the structure of the circumstances in which they are uttered.

Now poems and plays and novels have this in common with marriage ceremonies and trials, that their procedures are governed by quite elaborate and artificial conventions. These may be generic, rule-bound by the prescription of ancient practice or theory or even typographical. Unlike the marriage ceremony or criminal proceedings, they require approximate, not exact, observance on the part of the persons involved. Any comparison between 'institutional' and literary activities will have to take this difference into account. Nevertheless there is a sense in which the illocutionary force of speech acts in imaginative literature is conventional according to this definition. When Archibald MacLeish asserted that a poem should not mean but be, and William Carlos Williams insisted that a poem is an 'object' that 'formally presents its case and its meaning by the very form it assumes', they were thinking along these lines, though their conception of meaning was excessively restrictive and would not have passed muster with Strawson and Grice. The suggestion that the language of literature might be conventional in this relatively restricted sense, as well as in the less restricted sense already referred to, brings to our attention once again the superfluity of summoning ghostly agencies – intentional or otherwise – to account for the way speech acts are supposed to be governed. Again, whatever illocutionary force the words spoken or written possess – whether or not this is an intentional force – is inseparable from the fact of their being spoken in the sequence of speech acts that is being performed.

For the intentional aspects of the literary speech act we need to return to Strawson. He is not satisfied that an illocutionary act is capable of being made explicit with the help of an explicitly performative formula which would be called 'conventional' solely on the grounds that it has that capability. Accordingly he restates what is happening when such an action is performed in terms of intention. The restatement, which has been quoted above (p. 39), is as complicated as the intention it describes is 'complex'. But it does have the merit of emphasizing the importance of the exchange of understanding between speaker and audience, an understanding Strawson expresses in terms of intending to produce, and recognizing an intention to produce, a particular response on the part of both of them. The use of the explicit performative form is one significant means the speaker has at his disposal to make his intention clear enough to be recognized. But even where it is not used in an illocutionary act, its force is still present implicitly in the utterance. It is impossible to conceive of a speech act that doesn't in some sense contain an intentional illocutionary force. And in any one speech act that intentional force will apply to some particular matter to which it reaches out to secure a specific response. In a work of literature this will also appear to be the case, and in so far as we are concerned with the illocutionary force of the statement – its description recast in terms of an intentional understanding – the appearance will also be the reality. But because the sense and reference of the locutions of a literary text are to be understood metaphorically, this again makes a great deal of difference to just *what* it is that the writer is producing and the audience is recognizing. It is more likely to be something close to the fact of intentional force in and for itself than anything to do with the direction, in the sense of specific aims and goals, that in the normal course of events we would expect the illocutionary act to point towards.[66]

Poems are made out of speech acts that display a special kind of illocutionary force which is both conventional and intentional. Also poems are acts, the sum total of the speech acts they comprise. Perhaps we can find out more about them, and the place of intention in them, by considering them as acts – acts which have the special characteristic of being constructed out of speech or writing, but with enough in common with other kinds of human activity to make examination of them in this light pay a specifically literary critical dividend.

GOLDMAN'S ACTION PLANS

A good place to begin an account of contemporary theories of human action is A. I. Goldman's book on the subject.[67] For Goldman an act is an instance or occurrence at a particular time of what he calls an 'act-property'. It follows that to show what an act is like you have first to elucidate what an act-property is. For the purpose of my argument I shall simplify, as far as possible, Goldman's painstaking distinctions between acts, act-properties, act-types and act-tokens, referring instead to 'acts' as descriptions of a single way of interpreting 'what is happening' or 'what happened', and 'actions' as what underlies the sum total of descriptions that go to make up the character of a happening.

Goldman offers two complementary elucidations of an act-property, or what constitutes an action. The first one he derives from Anthony Kenny's distinction between three kinds of verbs, in chapter 8 of *Action, Emotion and Will*.[68] These are static verbs, performance verbs and activity verbs. Roughly, static verbs express states in which their subjects find themselves. Kenny offers as examples 'understand', 'know how', 'be blue' and 'be taller than'. 'Intend' is a static verb. All static verbs are incapable of being properly used in a present continuous tense. It makes no sense to say that someone is being taller than somebody else, or that something is being blue. It is easier to make the point with these 'be' verbs than the others, but it is still true that in spite of the idiomatic use of the phrase it makes no sense to talk about somebody understanding or intending as if he were performing a continuous action. These words run in parallel with other action words that are in some sense understood in the use of the static verb. Both performance verbs and activity verbs are action verbs, in the sense that they name actions. Performance verbs are distinguished from activity verbs by virtue of the fact that in any particular instance of their use the application of the continuous tense requires that the preterite is not applicable. Hence 'discover', 'grow up', 'build a house' and 'cut' are all performance verbs. A person who is discovering something has not discovered whatever that thing is. A person who is growing up has not grown up. Where an activity verb is used, on the other hand, the application of the continuous tense requires that the preterite *is* applicable. Kenny offers 'keep a secret', 'laugh', 'enjoy' and 'stroke' as examples. To be keeping a secret is to ensure that the secret has been kept. If you are stroking a cat, you have stroked it.

It isn't necessary for our purpose to worry about the difference between performance verbs and activity verbs, but it *is* important to distinguish between both of them and static verbs. So the terms I shall use here are the terms Goldman uses, which are active verbs and static verbs, or actions and states. To understand and to discover are both actions, and to intend to perform either of them is to apply to them an intentional state. To merely intend is not to do anything. 'An intention is not a momentary occurrence.... Intentions, like beliefs, are not always and necessarily the outcome of a process of thought or of a datable act of decision. They may, like beliefs, effortlessly form themselves in my mind without conscious and controlled deliberation.'[69] To do something, you have to perform an action, which might or might not be performed with the added force of an intention. For 'intending' to acquire the character of an action it must be associated with an instance of one or other type of action verb. When this happens, the manifestation of intentional force is entirely dependent on the way whatever activity is named by the action verb expresses itself in a specific act, or what Goldman calls an act-token.

Goldman's second method of elucidating actions uses the distinction Kenny has drawn between action verbs and static verbs to provide an account of the first in terms of descriptions of certain aspects of the second. These aspects are the intentional ones, what Goldman describes as notions of purpose, intention or reason. 'It has often been remarked', he says, 'that the distinguishing feature of human action is the element of purpose.... Doing something in order to achieve a certain end, or for a certain reason, or with a certain intention is central to the concept of a human doing.'[70] Later he acknowledges that by no means all human actions are intended. For example, insulting John, he says, doesn't have to be intentional. The important thing is that an act-property, as distinct from an instance of an act-property in the shape of an act-token, has to include the potential to be used intentionally in an act-token. However, in performing the act-token of insulting John I have to intend to do something else – perhaps amusing John, or insulting Peter – that happens to have the effect of insulting John. In other words, in the background of every act-token there must always be an intentional act-token that Goldman describes as 'basic'. And 'since all act-tokens must be somehow related to basic act-tokens, this imports the concept of purpose or intention into the heart of the concept of human action.'[71] Accordingly, any satisfactory description of a human action which is what Kenny describes as a performance or an activity (i.e. for Goldman, an act-token) must take into account the originating contributions made to it of the basic act-token which instantiates its act-property, and that basic act-token must be conceived as functioning in an intentional state. Most of the rest of the first half of Goldman's book demonstrates the kinds of relations that hold between these intentional basic act-tokens and other act-tokens that depend on them.

Relations between different act-tokens that are dependent on a single basic act-token hold together by what Goldman calls 'level-generation'. I shall try to explain what he means by this by offering an example of a fairly simple action that comprises a series of acts (act-tokens – from now on we can forgo the use of this rather clumsy technical expression) formed by level-generation from a single basic act. The example is figure 9 of *A Theory of Human Action*,[72] and it is used to clarify what is going on in the statement. 'By signalling for a turn I convinced my driving examiner that I am not a competent driver, where it so happened that I also got my hand wet, deterred a pedestrian from crossing the street, and convinced my examiner that no bearded men are competent drivers. (I am bearded.)' The figure, what Goldman calls an 'act-tree', makes clear what is the relation that holds between *both* the originating basic act and the intentional acts that are generated by it and by one another, *and* non-intentional acts generated in one way or another from the ones that were generated from the basic act. All these acts are described as being level-generated because they are performed during the same interval of time. (They are all happening together.) Even so, they occupy different positions in the act-tree in order to represent the different stages they have arrived at in the non-temporal 'process' of generation of the meaningful content of the action. My action plan (Goldman's phrase) was to convince my examiner that I am a competent driver, and I thought I was doing this by signalling for a turn. The fact that I did it at the wrong time had the effect of convincing my examiner that I was not a competent driver.

Extending my arm is the basic act I perform in doing all of the other things that appear in the diagram. It is an intentional act, related to the other intentional acts of extending my arm out of the car window and signalling for a turn. Let us for a moment simplify the action by substituting 'Convincing my examiner that I am a competent driver' for 'Convincing my examiner that I am not a

The act-tree is as follows:

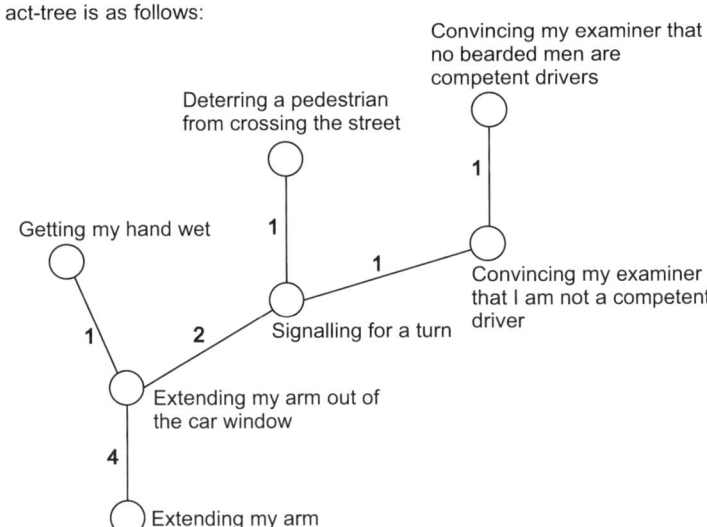

competent driver'. This act then becomes a fourth act that is intentionally generated by my extending my arm. All of these four level-generated acts occur simultaneously and are aspects of a single intentional action. If we replace the fourth in the series with the original 'convincing the examiner that I am not a competent driver', this becomes an unintentional act that is nevertheless causally related to the intentional act of signalling for a turn that diagrammatically preceded it, and convincing my examiner that no bearded men are competent drivers is in the same way causally related to convincing my examiner that I am not a competent driver. The same applies to the other two causally determined, level-generated, but unintentional acts of getting my hand wet and deterring a pedestrian from crossing the street. Both intentional and unintentional acts within the same act-tree, then, can be causally generated.

In the diagram (1) represents causal generation, and (2) represents conventional generation, which is characterized by 'the existence of rules, conventions, and social practices in virtue of which an act can be ascribed to an agent, given his performance of another act'. In the example, the other act is the extending my arm out of the window, which conventionally generates the meaning that I am signalling for a turn. (4) represents augmentation generation, where an act is generated by the augmentation or addition of some relevant fact or circumstance. Here the fact is the presence of the car window through which I extend my arm. The missing category of level-generation, (3), not represented in the diagram, is simple generation, which is 'like conventional generation minus the rules', i.e. 'the existence of certain circumstances, conjoined with the performance of [a generating act] ensures that the agent has performed [another act]'. If we were to add to the driving test diagram at 'signalling for a turn' a diagonal line with the caption 'signalling for a turn more vigorously than the previous examinee did', we would have an example of simple generation.[73]

Goldman is careful not to overstate the way in which and the extent to which it is permissible to describe a whole action plan as intentional or purposive, though he insists that the basic act that logically precedes all the others must be intentional. This is how he guards against a too comprehensively intentional description of an action:

> At first glance, it seems possible to define the notion of level-generation in terms of means and ends, or some other teleological notion. There are many cases in which a lower act is done *in order to* do a higher act or as a *means* to performing a higher act. John moves his queen to king-knight-seven in order to checkmate his opponent. S flips the switch in order to turn on the light. Unfortunately, teleological notions will not provide us with an adequate definition of generation. For while many instances of level-generation are accompanied with an appropriate purpose, intent or goal of the agent, many are not. Act A often generates Act A', though the agent does not do A *in order to* do A'. John's checkmating his opponent causally generates his giving his opponent a heart attack. John's moving his hand causally generates his fighting away a fly, but John does not move his hand with the *intent* of fighting away a fly. (38–9)

In just the same way, I signalled for a turn in order to convince my examiner that I was a competent driver, but what I did in fact was to convince him of the opposite. Both acts were causally generated, but only one of them was intentional. Even so, it is a fact of life that when we are aware of things happening in the normal course of events – which means when we are aware of the accumulation of co-temporal acts) into a series – we cannot help attributing notions of intention and purpose to the series that has been accumulated.

'Sequences of acts are important in identifying an agent's purpose because one often engages in an extended progress of actions all aimed at a single goal' (117).

What difference, if any, do we find when we apply this method of level-generational explanation to acts of writing? Here the rudimentary characteristic of disinterestedness is bound to make us redefine the notion of purposive or goal-directed behaviour that plays such an important part in actions like taking a driving test, playing chess and switching on a light that Goldman takes as recurrent illustrations of his theory.

The actions we perform in our everyday lives normally display symptoms of efficiency. An intentional state that accompanies the performance of an action gives an effective edge to that performance. To behave purposefully is to move towards the accomplishment of one's purpose in the most direct and uninterrupted way consistent with the fulfilment of all the requirements of the goal that has been set. To revert to Goldman's act-trees, the action plan constituting the 'trunk' of each of the trees is unaffected by the laterally generated 'branches' extending from it. The most successful action among the examples provided from the driving lesson was the one that 'developed' from extending my arm to convincing the examiner that I was a competent driver. This is because that particular co-temporal combination of events was all a part of my action plan, and my action plan was intentional. Getting my hand wet and deterring a pedestrian from crossing the road were merely causally level-generated actions that happened to accompany the basic act and the action plan that was generated by it. Convincing the examiner that I was not a competent driver and that no bearded men ever are competent drivers were expendable and unintended causally generated acts that displayed the ineffectiveness of my action plan.

An action that comes into being as a result of aesthetic considerations, however, displays an altogether different kind of relation to level-generated acts that 'fail' to belong to the action plan, and the basic act that underpins the structure of the whole action is to that extent less unambiguously basic. Unintentional level-generation acquires a value that it never has in practical activities like taking a driving test and switching on the light. I lift up my arm to flick the switch and thereby light up the room, and I also happen to disturb a fly. Disturbing a fly is just an accident (causally but unintentionally generated by my activity in switching on the light). But in a poem I imagine doing the same thing, and I record imagining doing the same thing, and the fly becomes a part of the poem. The unintentional activity of disturbing the fly retraces its path back to the 'branch' point in the action plan where it originated, and it achieves an intentional momentum retrospectively. This affects the character of the action plan. The same applies to level-generations that are simple, conventional and augmentational.

Goldman uses the phrase 'level-determinate' (59) to describe non-intentional acts that are nevertheless 'caused' by the agent's action plan. Disturbing a fly or getting a hand wet is a level-determinate act that fails to become part of my intention to switch on the light or signal for a turn. An example of the equivalent act in a poem is (to stay with the same image) Donne's importing into the second stanza of 'The Canonisation' the image of a fly, and then discovering that its figurative possibilities develop into the creation of another image – of moths getting their wings singed when they are drawn into the light of a taper. This in turn develops the possibilities of Donne's meditation on the dangers and delights of love, in a way unpredicted and unpredictable from what we had understood to be his initial poetic 'action plan'. So in poetry, level-determinate acts can accumulate an intentional significance that is belied by their origin in an act of choice that needn't and probably didn't include that particular intention. Who knows how that other little fly, in Blake's poem, first presented itself to his 'thoughtless' hand, or how 'blind' was the hand that the poet imagines brushing his own wing at the close?

Because the poet's premium is not on efficiency and directedness, the way different parts of the action behave in relation to each other in a poem or a play is not the same as the way they do this in other kinds of activity. It isn't possible to assess a poem in terms of the successful deployment of whatever intentional force is required to generate the various levels of activity in the original action plan.

This is easiest to see and explain in poetry, where the figurative character of the language used is most transparent. The value of a word in a line of poetry is assessed both in relation to the position it occupies in the metonymic forward movement in the line, and in relation to the metaphorical movement among all the rejected alternatives that fade away from it across the horizontal axis. The metonymic axis is the trunk of Goldman's act tree, the horizontal diagram of the action plan. The metaphorical axis is the diagonal 'sequence' of unintended level-generated acts that refuse to remain apart from the evolving structure of the intentional action. Keats wrote that the essential element of poetry is metaphor, and he was right. But 'metaphor' is not just using images for purposes of substitution or comparison. It can include any feature of language that works against the grain of efficient coincidence of the logical aspects of intentional level-generation. One way of construing what is happening during the writing and reading of a poem is to say that the character of a basic intentional act is enriched by the poet's discovery of the intentional force attributable to level-generated acts that might be causally related to it, but that are not, by virtue of that fact, intentionally charged. The way the action plan derived from the basic intentional act is made to absorb that intentional charge accounts for characteristics of imaginative fictions that mark them off from other kinds of writing. An activity that has been inaugurated for the effective realization of goals is transformed into one that takes cognizance of goals only as an opportunity for putting its intentional potential into play.

CONCLUSIONS

Taken together, the insights afforded by contemporary philosophers of language, action and meaning suggest that Wimsatt and Beardsley were mistaken in insisting on the privacy, and the unavailability to public inspection, of intentional acts. Intention has to be defined in a way that allows it some measure of publicity, at the same time as it is identified as a force that accompanies action rather than either a hidden cause of action or an action in its own right. If intention remains something entirely subjective, a Foucaultian wrinkle[74] in our knowledge, or an impulse in the author's consciousness which in some way or other undergoes a complete transformation before it enters the text he produces, then it does become a non-entity from the point of view of interpretation, let alone critical evaluation.

The argument that follows seeks to show that it is possible to reinstate intention as a property falling within the public domain. To do this it will be necessary to follow through some of the ideas about intention suggested by contemporary philosophers. We shall define intention very differently from the way Wimsatt and Beardsley did, though attaching to it an idea of personal agency foreign to such manifestations of the contemporary *Zeitgeist* as those displayed in Thom Gunn's poetry and in the theories of the poststructuralists. It should be possible to construct a theory of literature that acknowledges the basic soundness of the structuralists' concern with the grammar of literary discourse, while demonstrating that their apparently varied techniques of analysis have in common a single but fatal misconception of what this concern entails. We need a theory of literature that builds into the foundations of mental activity a rudimentary intentional force. At some level of understanding this must be brought into relation with a concept of purposiveness that inheres in our experience of the world as it is replicated in literary texts. The structuralists' concern with formal aspects of language and consciousness; Holloway's idea of the re-animation in apprehension of the twin consciousness of the writer in the act of writing and the reader in the act of reading; the philosophers' redefinition of intention as a force accompanying an action rather than an action in its own right, or a cause as an object of an action; and, as an aesthetic foundation, the notion of purposiveness without purpose that is at the heart of Kant's discussion of art and poetry – these are some of the approaches to the subject that I shall try to combine in the pages that follow.

World and consciousness are experienced as fitting together in the dynamic structurings of art. How this happens is obscure and mysterious. As Wittgenstein reflected, in one of those flashes of inspiration that light up his meditations on art:

> The 'necessity' with which the second idea succeeds the first. (The overture to 'Figaro'.) Nothing could be more idiotic that to say that is *'agreeable'* to hear one after the other – all the same, the paradigm according to which everything is *right* is obscure. It is the natural development. We gesture with our hands and are inclined to say: 'Of course!' – or we might compare the transition like the introduction of a new character in a story for instance, or a poem. This is how the piece fits into the world of our thoughts and feelings.[75]

But however obscure, there is no reason why the necessary succession of ideas should be totally hidden from observation. We can go some distance towards interpreting what it is that those gestures of the hands are expressing about *Figaro*.

How, then, are we to *show* this 'natural development', and how *show* its relation to authorial intention? It would be unwise to begin with intentions of *writing* when so much that is mysterious surrounds them. But what do intentions look like when they are expressed in the bolder, less introverted activities of life? To find out, we need to contemplate a picture of life that we have in common, and where better to find such a picture than in Shakespeare? If Shakespeare can't be trusted to provide us with a convincing picture of intention in action, who can? Let us start by observing a famous example of the dramatic realization of a bold intention: Macbeth's decision to murder the king in the first two acts of *Macbeth*.

2
Shakespeare

> Thoughts are no subjects,
> Intents but merely thoughts.
> > Shakespeare, *Measure for Measure*.

MACBETH AND WITTGENSTEIN

> . . . . the dark intent I bring
> > *Paradise Lost*, Book IX.

Before the murder of Duncan, Macbeth speculates 'If it were done when 'tis done, then 'twere well/It were done quickly' (*Macbeth*, I. vii. 1–2). With each new line, with each new image, Macbeth pushes himself towards the act of murder, then pulls himself back from it. The language of the speech is condensed almost to the point of unintelligibility. It is rare even for Shakespeare to lay bare the innermost workings of a man's mind with such precision and immediacy, and to achieve that effect he makes use of the full resources of his mature dramatic style.

The speech ends as follows:

> I have no spur
> To prick the sides of my intent, but only
> Vaulting ambition which o'erleaps itself,
> And falls on th'other.

The poverty of critical commentary on these lines of *Macbeth* is startling. To the best of my knowledge no critic has got far beyond Kenneth Muir's footnote in his Arden edition of the play, paraphrasing the lines as follows: 'I have no spur to stimulate my guilty intention except ambition – ambition which is like a too eager rider, who in vaulting in the saddle o'erleaps himself and falls on the other side of the horse.' Muir then summarizes the views of other scholars:

Hunter explains: 'lights on the opposite side of what was intended; that is, dishonour and wretchedness, instead of glory and felicity.' Wilson mentions that vaulting into one's saddle was a much admired feat. Grierson, following Stevens, suggests that 'Shakespeare may be thinking of a too furious rider who, leaping too high at an obstacle, clears it indeed, but falls on the other side.'[1]

In a study of *Macbeth* stretching to eight hundred pages, Marvin Rosenberg concludes that 'the whole [of these three lines] does not bear detailing'. He adds:

> the images are surreal: one spur pricks both sides of an impulse; *vaulting ambition* does not simply *o'erleap* to fit a convenient equestrian figure, but leaps over itself and falls, suggesting a grotesque somersault and crash – projecting the course of Macbeth's present inner process as well as his future career.[2]

That is all.

The third and fourth lines are very puzzling. It is easy to see why commentators have shied away from them. They contain two images, or pictures, linked together by the common application to riding a horse. The first finds the speaker mounted, but without spurs. The second finds him about to mount, but failing to do so as a result of over-eagerness issuing in miscalculation: he intends to mount into the saddle, but jumps too energetically and slips off the horse onto the other side. In the first image, the horse 'stands for' Macbeth's 'intent' (to kill the king). In the second it must surely stand for something else, something that comes between Macbeth and what he foresees he is about to accomplish – again, killing the king. So now 'killing the king' lacks an image. It is just something Macbeth realizes he is about to fall into. But what it is, what lies on the other side of the horse, so to speak, is vacuous and enigmatic. This must be the case if the vaulting is intended to be a vaulting into the saddle from a position on the ground. I favour this interpretation because it is the almost unanimous preference of Shakespeare's editors, from Cunningham to Muir,[3] and because the last three words of the speech, 'on th'other', seem to be an elliptical rendering of 'on the other side' – 'side' having been used in the first half of the sentence to denote the side of the 'intent'/horse image.

It could be interpreted differently. If we forget about the horseman on the ground at the end of the first clause, we can begin the second (at 'but') already mounted on the horse, and then make the horse stand for Macbeth's ambition. The horse, not Macbeth, then does the vaulting. With or without spurs, Macbeth (the rider) encourages the horse (his ambition) to 'o'erleap itself' in the sense of 'aim to do more than in the event it proves capable of doing', with the disastrous result – 'And falls on th'other'. The trouble with this interpretation is that the meaning of 'itself' is a little forced, and 'th'other', which no longer means 'the other side', becomes an even more mysterious non-entity than it was in the conventional reading. Its only advantage – a large one – is that it dispenses with the irrelevantly comic image of a rider jumping over the saddle of his horse and falling down on the other side of it. It also makes some sense of the order of the two parts of the sentence: man on horse → man on horse jumping, instead of man on horse → man about to get on horse and then failing to do so.

In this speech of Macbeth's Shakespeare is describing the way an intention comes into being, how it is formed. Many other events prompt Macbeth to think about killing the king, wonder whether he should kill the king, ask himself whether killing the king will produce this result or that, *etc etc*, to the point at which, here, or hereabouts, he comes to the brink of decision. Only Lady Macbeth's last tauntings ('letting "I dare not" wait upon "I would"') stand between his intention and its fulfilment. His conversation with Banquo does not deflect him, and the phantom dagger 'marshall'st me the way that I was going' – i.e. it enables rather than obstructs. The speech represents the hardening of Macbeth's weaker into his stronger intention. By the end of it, the intention is still not strong enough or firm enough. But after the dialogue with his wife, Macbeth remains in a state of decision indicated by his words 'I go, and it is done'. If at the time there should appear to be anything suspicious about the combination of the active and the passive voice in that statement, it is allayed by his next words to Lady Macbeth: 'I have done the deed.' Somehow the two parts of the horsemanship image have coalesced in the murder, and the spur of Macbeth's intent has brought about the vaulting ambition that has o'erleapt itself.

Shakespeare has described the last stage in the coming into being of an intention and its hardening into the real precipitant of an action. He has done so with a concentration of imagery, syntax and wordplay that draws attention to the extreme difficulty of representing this subject. The image of 'intent' solidifies (into a horse), then vaporizes ('other' what?), or slides into correspondence with another abstract noun ('ambition') which is shown to be only very partially a correspondence because of the complications introduced in this context by 'vaulting'. Sometimes it seems to be voluntary, purposive and deliberate; sometimes unconscious, involuntary and mechanical. At the moment of occurrence, intention is a much more complicated state of mind, or temperamental inclination, than in our less thoughtful moments we give it credit for being.

Shakespeare is working at three levels of intention in his enactment of Macbeth's will to murder. At the lowest level, we infer something about the behaviour of a human agent who is part of a metaphor. The rider intends to jump into the saddle but fails to do so. In fact 'rider', 'saddle' and 'fail' are not words that appear in the metaphorical statement. They are properties of the speech that we infer from the presence in it of 'vaulting' and 'falls'. This is happening in a context where horsemanship is clearly in the forefront of the author's mind and the word 'ambition' is acting as a kind of personification ('o'erleaps *itself*').

The middle level of intention applies to what I take to be a deliberate implied connection between the constituents of two successive sentences: the horsed cherubim and the vaulted and o'erleapt sides of the (horse). Here the inference about intention works differently, because the attribution of intention comes from a different source. In the lower-level example it was our assumption that if x is described as failing to do y, then x must have intended to do y, even if we weren't told that this was the case. The assumption is common to the author and ourselves, and is made from the same point of vantage and with the same degree of certainty.

It is in the nature of language that in places like this if it says one thing it must also imply another, and the writer and reader are equally bound to acknowledge the implication. In the middle-level example the reader is inferring that the author intended to produce a certain effect, on the grounds that the arrangement of the words in these last two sentences makes it probable they will register in his (the reader's) mind in the way that has been described. The only other inference that could have been drawn is that the combination of those particular words in the two sentences is accidental. This is perfectly possible. Whether the effect of Shakespeare's combining these words in this particular way is judged to have been deliberate – i.e. intentional – or accidental, will depend on the reader's inference about the intention at the third, or higher, level.

In the first and second acts of *Macbeth*, the illusion has been produced of a man converting an intention into a fact, an inner impulse into an effect in the world beyond his consciousness. But because it is an illusion, and not an event we witness from somewhere outside the perimeter of Macbeth's mind, we can see both what is going on in Macbeth's mind and 'what happens' – in the sense of a chain of events, and one event in particular, that is brought about as a result of Macbeth's thinking like that. We can see 'what happens' as something that would not have happened were it not for the fact that Macbeth's mind, which somehow included his intentions, worked the way it did. Shakespeare has exposed it to our view in his soliloquies and his conversations with Lady Macbeth.

The first thing we discover about Macbeth's state of mind (on his first appearance at I. iii) is that he is fearful. He has heard the prophecies, and he is fearful of their import. But Macbeth doesn't tell us this, and he doesn't tell himself this either. We rely on Banquo's interpretation of the 'start' he gives (51) before he 'seems rapt withal' (57). Being 'rapt', he says nothing between lines 47 and 70. According to Banquo, Macbeth 'seem(s) to fear/Things that do sound so fair' – the things being the prophecies. Before the sisters vanish, all Macbeth says about their prophecy that he will be king is that 'to be king/Stands not within the prospect of belief' (73–4). At this point in the play, then, there are only three things that bring together in our minds the idea of 'king' and the idea of 'Macbeth': the sisters' prophecy, Macbeth's dismissive attitude towards it and his fear (according to Banquo). There is not enough to go on here for an audience even to begin to wonder about Macbeth's intentions about being king. He appears to have none. His fear is unintelligible. In any case, Banquo might have been mistaken.

After Ross has entered and told Macbeth that Duncan has made him Thane of Cawdor, we are given our first opportunity to see directly what is going on in Macbeth's mind: 'Glamis, and Thane of Cawdor!/The greatest is behind' (116–17). Since this is an aside, the conventions of Jacobean dramatic writing dictate that what Macbeth speaks is true, or is believed by him to be true. Here the qualification is irrelevant. Macbeth truthfully puts it to himself that the greatest (i.e. the first two out of three) of the sisters' prophecies have come to pass. But if two of three things are 'behind', in the sense Macbeth means here, then the supposition must be that he is aware that the other one is 'before' him. All this means at present is that he is still not king. There is no suggestion about how he might become king. The only matter that remains unspoken is the tacit authority he grants the sisters by immediately associating Ross's information with their prophecy. But since only about three or four minutes' stage time have elapsed since the sisters made the prophecy, it would be odd if Macbeth hadn't formed that association. Banquo has done so too ('What! Can the devil speak true?' – 107). Banquo's lines about the 'instruments of darkness' seem to be for our benefit, not Macbeth's. What Macbeth says immediately afterwards need not imply that he was listening to Banquo. At the close of his soliloquy (126–42) Banquo again refers to Macbeth's being 'rapt'. Apart from 'I thank you gentlemen', spoken in order to clear Ross and Angus from the centre stage, Macbeth seems to have been in this condition for almost all of the time that has passed since Ross brought the good news. The soliloquy provides us with our first opportunity to enter Macbeth's mind for any length of time. This is where we can expect to find out whether he has any intentions that are sufficiently private not to be disclosed to Banquo, or to anyone else on the stage for that matter.

The first two lines of the soliloquy do no more than repeat the gist of the preceding aside (though with hindsight the phrase 'swelling Act' might seem to carry more weight than anything we found there at the time). Thereafter, the reference to 'supernatural soliciting' for the first time suggests to us, though not to Macbeth, that Macbeth has interpreted what the sisters said as an 'incitement' or 'allurement' to do something the 'good' or 'ill' of which he cannot at present determine. Macbeth's reflections are wholly taken up with the moral good or evil represented by the sisters' prophecy, and of course this is something which we take an interest in too. In other words the good, the ill and the truthfulness of the prophecies so far are the objects of Macbeth's deliberations. The 'supernatural soliciting' is not. It is just a form of words he uses to establish the direction of his thoughts. For the audience, though, that phrase *is* an object of attention, and it is a very important one. The fact that Macbeth uses the word 'soliciting' is an indication of his state of mind, and it might also be an indication of what his intentions are or will be. This is so because we know there is no need for Macbeth to interpret what the sisters said as an incitement (to do something). Prophecies are descriptions of future events. They are not (or need not be) promptings to bring those future events about. So the word 'soliciting' tells us something both about Macbeth's cast of mind, *and* what might be the formation in it of an intention that Macbeth himself is not aware of at this stage.

He becomes aware of it when he describes as evil his yielding 'to that suggestion/Whose horrid image doth unfix my hair...' (134–5), where 'suggestion' carries the same implication of incitement that 'soliciting' carried a few seconds before. But this time the 'suggestion' is something that Macbeth contemplates. Since he chooses the word to define what it is that he 'yields to', he cannot be unaware of what it means, and the references to a horrid image unfixing his hair and making his heart knock against his ribs show this is the case. But it is still 'that suggestion', something prompting evil action, but leaving the precise nature of it undisclosed. The language of 'horrible imaginings' remains vague, though the 'Present fears' they are 'more than' throws back a suggestive connection to the fear Banquo mentioned at the beginning of the scene. It helps to tie in Macbeth's present (so far inarticulate) intentions with what might be construed as intentions, or the coming into being of intentions, he allowed into his mind soon after the sisters had spoken. But now he thrusts aside words like 'suggestion', 'horrid image' and 'horrible imaginings', that have disguised his intentions even in the act of forming them, and comes out with the more specific word that for the first time exposes to us what his intention is, or at any rate what it might be (what it is *in thought*): 'My thought, whose murder yet is but fantastical.' He is imagining murder, which is itself an evil; but he has not arrived at the point where he could say of himself, or we could say of him: 'Macbeth intends to kill the king.' Instead, 'function is smothered in surmise': the intention to murder is not ready to detach itself from other kinds of imagined activity – like waiting for the king to die, or horrifiedly trying to shake off any prompting to assist his dying. What Macbeth says to himself here can be read in two ways: the function (i.e. imagined murder of Duncan) is smothered in surmise (other wonderings that prevent the idea of murdering Duncan from pulling free and becoming an intention rather than an act which is horrifyingly seen to be possible); *or* the function (the whole mind and imagination of Macbeth) is smothered in surmise (of what it would be like to murder Duncan, if that possibility were to lodge itself in the mind as a single intention, rather than as one among the many possibilities that at present occupy it).

Important points to notice here, at the end of the first scene in which Macbeth appears, are: (a) no one else in the play understands what Macbeth's intentions are; (b) Macbeth's intentions are so confused, contradictory and inarticulately formulated that he himself cannot be said to be fully aware of them; and (c) as a result of the privileged insight it has acquired into Macbeth's mind (through the soliloquy), the audience knows more about what Macbeth intends to do, not merely than the other characters on the stage, but than Macbeth himself. What Macbeth brings forward as the subject of his own attention becomes an object of the audience's attention as well. The way Macbeth goes about bringing forth such objects of attention, the manner in which his mind shapes its own habits of inquiry, also becomes an object of the audience's attention. A subjective process of Macbeth's mind becomes an object of critical moral attention in our own.

The fact that Banquo described Macbeth as 'rapt' ('Look how our partner's rapt') after the first soliloquy, as well as some time before it (77), suggests that he is in a sort of hypnotic trance during the period between meeting the sisters and speculating here about the way the first of their prophecies has come true. This tallies with what the words imply about his behaviour on the stage. Most of what he speaks, apart from the soliloquy, is in the form of asides. The rest is an address to the sisters ('Stay, you imperfect speakers!' – 70ff.); a musing sort of conversation with Banquo ('Your children shall be kings...' – 86 *etc*); and brief formal courtesies ('Thanks for your pains' – 117). Macbeth could have spoken any or all of this from within the same 'rapt' frame of mind in which we found him in the asides and the soliloquy. After Banquo's last 'rapt' description, Macbeth speaks two brief asides, which cause his companion to hint that he is behaving strangely: 'New honours come upon him/Like our strange garments, cleave not to their mould/But with the aid of use.' Then Macbeth seems suddenly to come to his senses: 'My dull brain has wrought with things forgotten' (149–50).

As his words to Banquo a second or two later reveal, Macbeth has not forgotten the prophecies ('Think upon what hath chanc'd'). So he must be referring to the 'suggestion' and 'fantastical murder' of the soliloquy. Either that, or thoughts of murder he had entertained before the play began and that the sisters' prophecies have brought to mind again. This would explain how Macbeth could at the same time have remembered and forgotten whatever the 'things' were. They *had* been forgotten, but *now* he has remembered them. The 'supernatural soliciting' speech conveys a strong impression of being a response to a frighteningly new experience (of contemplating – if that is the right word – the murder of Duncan), just as the 'If it were done' speech conveys a strong impression of the mind's having made such a response and being on the brink of acting accordingly. In the one soliloquy, an evil prompting struggles to the brink of consciousness and forms an imaginary intention ('whose murder yet is but fantastical'). In the other, the 'intent' is fully conscious, imaginary only in the sense that it has not yet expressed itself in the appropriate physical act. Macbeth's intention, weakly present in the wording of the conditional clause at the opening and more strongly in the enunciation of the riding image at the end, hesitantly but definitely becomes a perfectly articulated mental state. From the outset Macbeth knows what it is. There are no unspecific preliminaries to the articulation of the word 'assassination' in the second line – as there were to that of the word 'murder' in the earlier soliloquy. The triple repetition of 'done' at 1–2 brings vividly to mind the idea of 'deed' as 'murder' which is perceptible elsewhere in these first two acts of the play.

Macbeth's intention is fully formed at the end of I. vii, after the close of the 'If it were done' soliloquy, and after Lady Macbeth's 'screw your courage to the sticking place' speech that follows it (at 59ff.). 'I am settled', he says, 'and bend up/Each corporal agent to this terrible feat.' That, surely, is an expression of intention stripped of all the ambiguities and confusions that have prepared its way. Like Martin, the chief player in Barry Unsworth's *Morality Play*, Macbeth 'had passed from notion to intention to strategy as if there were between them no curtain, nor even a screen of mist'.[4]

'What is the natural expression of an intention?' asks Wittgenstein. 'Look at a cat when it stalks a bird; or a beast when it wants to escape.'[5] Macbeth is like a cat stalking a bird at the end of I. vii, and at the end of the next scene (II. i) when, on the way to Duncan's chamber, he clutches at the phantasmal dagger. The whole of this speech (33–64) brilliantly objectifies Macbeth's experience of intention. The real and imagined daggers – 'such an instrument I was to use' – represent exactly the same thing, his murderous intent, as does every word he utters as he strides towards the realization of his 'design'. The intention by now is fully conscious, voluntary and deliberate. It is an objective thing which looks like Macbeth's movements, feels like his grasp of the real dagger which the phantasmal dagger prompts him to draw, and sounds like the words he speaks – which describe precisely and with great vividness the activities (of clutching, drawing, pacing, striding, even breathing) he performs as he makes his way towards the sleeping Duncan. 'I go, and it is done' is the point at which the intent to murder crosses over into the act of murdering. Intention has as good as issued in realization. The cat has pounced, the bird is caught.

In *Philosophical Investigations* Wittgenstein says little about intention,[6] rather more about deliberation – which is closely related to it. (There is a progress from voluntary to deliberate to intentional action in Macbeth's appreciation of his predicament.) When Wittgenstein refers to deliberate actions, he seems to be insisting on their objective appearance. For instance, at I, para. 174 we read the following:

> Ask yourself how you draw a line parallel to a given one 'with deliberation' – and another time, with deliberation, one at an angle to it. What is the experience of deliberation? Here a particular look, a gesture, at once occur to you – and then you would like to say: 'And it just is a *particular* inner experience.' (And that is, of course, to add nothing.)

It is the same with the cat stalking the bird. Here too 'a particular look, a gesture, occur to you'; but in this case we are less likely to 'add nothing' by referring to a 'particular inner experience', because we don't usually assume we know what it feels like to be a cat. So it is with Macbeth. The convention of the soliloquy allows us the illusion of penetrating Macbeth's mind during the period when he is formulating the intention to murder Duncan. By a conventional literary process of turning-inside-out, Macbeth's *'particular* inner experience' is transformed into an outer manifestation of experience communicated in a public language. These are verbal equivalents of the 'particular look, a gesture...', to which nothing need be added. The expression of deliberation, very close to the expression of intention, either is public (the stage activity, the fact of

Macbeth's speaking to himself) or it is transformed into something that is public by Shakespeare's creation of a language for Macbeth to speak which, though not actually a part of the (illusorily) objective business of the stage, reinforces that business symbolically.

Wittgenstein moves on to the issue of intention itself at para. 247: '"Only you can know if you had that intention." One might tell someone this when one was explaining the meaning of the word "intention" to him. For then it means: *that* is how we use it.' At first this seems to be implying something different from what was implied in the note on deliberation. Surely intention must be private if only you can know if you had it? On closer inspection, however, it transpires that what Wittgenstein is doing here is exposing a peculiarity of the language game that concerns *talking about* intention. Really, intention as an inner movement, or prompting, is unknowable even to the person who says of it – interchangeably – that he had it and that he knows he had it. 'The reason for calling whatever experiences we have "experiences of intending" (the reason for seeing them *as* experiences of intending) is that they go with intending. If the same experiences went with "wishing", then they would be "experiences of wishing".'[7] As Wittgenstein says in brackets after the sentence about how we use the word: '(And here "knows" means that the expression of uncertainty is senseless.)' So in *Macbeth* I. vii, Macbeth doesn't 'know' he intends to murder the king. The intention to murder him comes into being as a result of another feature of Macbeth's mental activities – some of which have already been implied, others of which still need to be explained. We know of Macbeth's intention through his activity – including his verbal activity, which in turn includes verbal activity that 'Macbeth' as a dramatic vehicle is responsible for but that Macbeth as a character cannot be said to 'know'.

The most obvious way we know about Macbeth's intention is through its realization in the murder. In *Philosophical Investigations* Wittgenstein is for the most part talking about intentions which are not realized. At para. 645, for example, he discusses the characteristics of someone's saying 'For a moment I meant to . . .' He describes them as follows: 'That is, I had a particular feeling, an inner experience; and I remember it *quite precisely*. Then the "inner experience" of intending seems to vanish again. instead one remembers thoughts, feelings, movements, and also connexions with earlier situations.' This seems to be an accurate description of how we witness and how Macbeth experiences his intention to murder Duncan – though in his case the intention is further validated by the fact of the murder, i.e. not 'Letting "I dare not" wait upon "I would"' (I. vii. 44). It is also complicated by the fact of the murder, because that fact tends to crystallize the intention to murder in the audience's mind, and therefore bestows on it the status of a public activity. In fact it has that status only in so far as it is expressed as the cat expresses it when it stalks a bird, and, as we have seen, in the Macbeth parallel the cat stalking the bird is something equally externalized in words and pictures.

Wittgenstein has performed a useful service in drawing attention to 'meaning to' as a substitute for 'intending'. The phrase emphasizes the forward-looking character of intending. An intention has to be a movement of the mind or will towards the realization of an end that remains hypothetical until the intention is realized. A man who is intending something is therefore drawn towards bringing about a state of affairs that exercises a power over him from a hypothetical future state of his own imagining. That is one reason why he can express, but cannot know, his intention. For at the moment(s) of intending, his urge to know is directed outwards at the hypothetical future state, not inwards at the mental act that imagines the future state. The mind can't simultaneously reflect on its own operations and keep steady before it the future state towards which the will, in an intentional act or acts, prompts it. That is why the word 'intent' first appears at the end of Macbeth's first long soliloquy. The horse is to move forward after its sides have been pricked. But even in a metaphorical sense, it fails to register as the clear image of a horse in Macbeth's nightmare vision. Remember the point made about the illogical move from the image of a horse being ridden to that of a horse being vaulted upon (or over). When Macbeth introduces the image of the spur, he uses it to qualify the impetus of his intention, driving the horse forward into the realization of his hopes. But when he halts the imaginary activity of riding to focus attention on the spur as a representative symbol, he identifies it with 'vaulting ambition'. It is as if the only spur he had was that, or as if he didn't have a spur at all, but, instead ('only' = 'instead'), 'vaulting ambition'. The move from 'spur only in this sense...' to 'spur not at all, but instead...' is accounted for by a move from enacting how something is to come about, to reflecting why he wants it to come about. R. S. Peters describes a motive as 'an emotively charged reason...Motives...refer not only to the goal towards which behaviour is directed but also to emotional states which set it off'.[8] 'Ambition' is a motive, not an intention. It exists prior to the forming of an intention, not as some sort of end to which the intention is directed. In Aristotelian terms ambition is an efficient, not a final cause of the intention.

OTHELLO AND ANSCOMBE

Our thoughts are ours, their ends none of our own.
Hamlet.

We are in a better position now to understand the confusion Wimsatt and Beardsley betray in their essay on the 'Intentional fallacy'. It arises out of the fact that what they are discussing are the motivated not the intentional characteristics of an action. Motive and intention are very different things, though they are both likely to be constituents of a single sequence of mental events. An intention can easily remain one of those 'I meant to...' states of mind to which Wittgenstein refers. As such it will be an inner experience, a particular feeling, which dissolves in the mind before anything is actually done about it. On the other hand it may be, and often is, expressed in some form of action. Although the intention itself is bound to remain unknown – in the sense that there is no difference between knowing it and having it, i.e. intending – its expression in action will not. And from that expression the nature of the intention itself can be inferred. In this sense the expression of intending will look like Wittgenstein's experience of deliberation, because both involve doing something – doing something intentionally, doing something deliberately. They have an object in the sense that they prompt the mind to create the image of what the effect of their doing it would be – and this doing would be achieved by the successful realization of an intention, or of activity following deliberation. But the link between motive and intention, or between motive and the expression of intention in an action, might be very mysterious indeed, and not at all obvious. This is because of the 'prior' position of motive in the sequence: motive → intention → expression of intention → action. Linguistic philosophers use the phrase 'intentional action' to emphasize the identifiable connection between what is done and the intention with which it is done. A phrase like 'motivated intention' would by much more difficult to apply in many cases where it is not possible to explain with any degree of certainty the connection between how something is done and why it is done. There are occasions, though, when the link between intention and intentional action is itself weakened by the presence of uncertainties in the prior link between motive and intention. This creates a complication. It is a complication which we are familiar with in literature, and therefore in Shakespeare.

Probably the most highly developed manifestation of it is in *Othello*, and concerns the character of Iago's motives in relation to the character of his intentions. What is the link between what Iago intends to do and what he does? What, in fact, does he do? And what are his motives for doing it?

The only simple answer to any of these three questions is to the second. In chronological order, Iago effects the disgrace of Cassio, creates in Othello's mind a jealousy of Cassio's supposed relationship with Desdemona, and through that jealousy drives Othello to murder Desdemona and to kill himself. In spite of sophisticated (and possibly correct) critical theories about Othello's propensity to jealousy, and in spite of the fact that Othello kills himself because of what he discovers is the truth about the situation, not because of what he fails to discover is a lie about it, it is surely reasonable to claim that Iago bears the primary responsibility for these events. There is no difficulty about this. The difficulty arises when we press the case against Iago further, and claim to identify these events, for which Iago bears the responsibility, with the satisfaction or the realization of his intentions.

To speak of Iago's intentions so soon after speaking of Macbeth's is to encounter an unexpected difficulty. At the opening of *Macbeth*, Macbeth has no intentions relating to Duncan that are in any sense significant. He is a loyal subject who has just played an important part in the defeat of the king's enemies. Unless we pay more heed than we should to one possible interpretation of Lady Macbeth's speech at I. vii. 47–52, we are not to suppose that Macbeth entertains any ambition to seize the throne of Scotland. Then the unexpected happens. He meets the sisters on the blasted heath, hears their prophecies, and his mind begins to turn to the prospect of rebellion and murder. At the opening of *Othello*, though, it is difficult to tell whether or not Iago already has intentions regarding Othello. Two events have already occurred which have a definite bearing on his attitude to Othello: he has been passed over for the lieutenantship in favour of Cassio; and Othello has secretly married Desdemona. It is possible to imagine Iago taking in these facts and weighing their significance, or simply experiencing whatever emotions seem appropriate to their having occurred – anger and envy about the first of them, disgust and jealousy about the second. In both of these cases the matter of intention, what Iago means to do about them, will not be immediately relevant. There will be time to watch him formulating his intentions later in the play, as Macbeth does in the first two Acts of *Macbeth*. It happens, though, that Iago doesn't have the luxury of time to do this, because at the very moment the play opens more than his own fortunes depend upon what has come to pass. From the start, Roderigo tries to take the initiative. Not having gained the lieutenantship, Iago can be seen to be less well placed than he had hoped to procure Desdemona for Roderigo, and the wedding, which Roderigo has just this moment heard about, makes it look as if Iago has not been playing fair by him all along. So Iago either has to form intentions of what to do about this state of affairs right away, or he has to pretend to do so, in order to convince Roderigo that he knows what he intends to do. The trouble is that since Roderigo and Iago are alone on the stage in I. i, the audience have no one but Iago to rely on to tell them what he is intending to do. So how are they to discriminate between his real and his professed intentions? It is a problem to which the audience will address themselves throughout the play.

It is a problem because, although the play tells us what Iago does, we are perplexed about what his intentions are in doing it. This is because our assessment of his motives fails to correspond with our perception of what they have led him to do, and the link between the motive and the activity must be the intention. Iago must have an intention, or intentions. And there must be some kind of reasonable connection between motive, intention and action (and the effects of action). In Iago's case, however, there is a special pertinence in Carl Ginet's view that 'side effects of an intended action that the agent *expects*... need not... be effects the agent *intends* to bring about, but they are... effects (s)he brings about intentionally.'[9] With what degree of accuracy, then, can each of the three principal events Iago brings about be described as the deliberate realization of an intentional act which springs from the motives he articulates from time to time as the plot develops?

The opening scene between Roderigo and Iago (to line 74) puts difficulties in our path straightaway. What are they talking about? What is it Iago should 'never tell' Roderigo? What is 'this' (3) that Iago knows about but that Roderigo did not? We don't find out until line 68 at the earliest, when Iago's reference to 'calling up her father' introduces Desdemona (not by name until I. ii, but we know she is Brabantio's daughter by I. i. 78, and suspect she is the Moor's newly wedded wife at about line 73). After Roderigo's mysterious complaint at 103, Iago takes up two or three minutes' playing time complaining about the Moor's appointment of Cassio as his lieutenant. It is not immediately obvious why Iago should be saying this to Roderigo, whose interests lie in quite another direction. It is true that Iago is making excuses for his failure to procure Desdemona for Roderigo. But the time Iago spends making the excuse, and the time he doesn't spend revealing what it is he needs the excuse for, requires a lot of explaining – especially in an opening scene when the audience knows nothing, not even the names of the 'him' and, eventually, 'her', who keep on being mentioned.

By the end of this scene, having arrived at Brabantio's house, we may not know exactly who is who, but we do know two things about Iago that sound important.

The first is that he is consumed with anger at Cassio's recent appointment. Anger, even fury, is Iago's most prominent characteristic here. We suspect it might provide a motive for whatever he will do later on, but since we have no idea at this stage what he *is* going to do, this must remain for the time being merely a suspicion. Then, after his long speech about the Cassio lieutenantship, he expresses an intention: 'I follow him to serve my turn upon him' (42). This is vague, but does give us an early indication of the difference between Iago's apparent 'honesty' and his actual duplicity. It also shows us that he is going to do something about it, that he is disguising his real intentions in a cloak of unselfish loyalty to Othello. But what his real intentions are, he doesn't say.

This brings us to a second important fact we discover about Iago here. Through what he calls upon Roderigo to do we can infer what his own intentions are, in a general way. He tells Roderigo that 'though that his [the Moor's] joy be joy' he must 'Yet throw such chances of vexation on't/ As it may lose some colour' (72–4). Coming so suddenly after Iago's confession that he follows the Moor 'to serve my turn upon him', this command encourages us to forge a link between the general statement of intention there, and the slightly more specific statement of intention (about what, in this instance, he wants Roderigo to do) here.

In the first half of the scene Cassio seems to be uppermost in Iago's mind. Therefore the shift in direction to Othello at 'I follow him', confirmed by the speech about 'chances of vexation', is significant, deflecting the audience's attention away from Cassio and on to Othello for the first time as the principal object of whatever intentions Iago has in mind. Furthermore these intentions appear to have nothing directly to do with Othello and Cassio. Instead, they have to do with Othello and Desdemona. Indeed it is only at this point we begin to realize that Othello's relationship with Desdemona was what Roderigo was talking about at the very beginning of the play. So how are we to interpret Iago's eventually getting round to this subject after spending so long talking about Cassio? Is it because he has succeeded in lulling Roderigo's suspicions? (I have already discounted this as a primary motive for Iago's obsession with Cassio.) Or is it because Iago has said all he wants to say about Cassio and is now moving quite naturally and sincerely (in his way) on to a second and, as it happens, equally infuriating aspect of his service under Othello? Or is it just that the play demands a reference to Othello's marriage in order to clear up the mystery of Roderigo's first three lines? It might well be that the second explanation prompts the third: Iago is as much preoccupied with the Othello/Desdemona marriage as he is with the Cassio lieutenantship, and some stage business (the arrival at Brabantio's house) has to take place to divert his attention from the one obsession to the other. However, they are both obsessions, and they are both equally important contents of Iago's mind at the opening of the play.

They are equally important, but are they connected? That is where the difficulty about the relation between motive and intention arises. Iago's contempt for Cassio, added to Cassio's promotion, provides a motive for Iago to act – against Othello or Cassio or both. But his invention of the idea of 'chances of vexation', applied to Othello and Desdemona, is more like a statement of intention, albeit couched in the form of a command to Roderigo. Part of the difficulty is removed if we assume that all Iago has in mind is what his words immediately refer to, i.e. that he can't actually remove Othello's joy in Desdemona, but that he can take some of the gilt off the gingerbread by having Brabantio make a nuisance of himself at the Senate House. However, in view of what happens later than I. iii, I think we can reasonably assume that something more is beginning to fester in

Iago's mind than can be explained by this imputation of an intention of limited scope. How he will act, though, is still a mystery – no doubt as much to Iago as to us. What it has to do with Cassio's promotion is equally mysterious, except that Othello is involved in both the promotion and the marriage. This exception isn't enough, though, to account for the passion that goes into the expression of Iago's attitude towards Othello/Cassio and of his intentions about Othello/Desdemona. The first is evident in this scene, the second is prepared for at 70–4, but becomes more obvious a little later (at I. iii, in his conversation with Roderigo before the soliloquy).

As early as the first half of I. i there is a pecular discrepancy between Iago's feelings and his possible motives on the one hand, and his intentions on the other. The feelings are strong (though it turns out that we have an incomplete account of them here). The intentions are vague, but not so vague as to conceal the gap between what they are and the feelings from which we assume they must spring. Given the oddities and puzzles of this first scene, we are eager to hear Iago's own account of what it is that drives him to intend anything at all, and what, in greater detail, his intention is. It is in his soliloquy that we expect to discover the link between them.

> Thus do I ever make my fool my purse:
> For I mine own gain'd knowledge should profane
> If I would time expend with such a snipe,
> But for my sport and profit: I hate the Moor;
> And it is thought abroad, that 'twixt my sheets
> He's done my office; I know not if't be true;
> Yet I, for mere suspicion in that kind,
> Will do as if for surety. He holds me well;
> The better shall my purpose work on him.
> Cassio's a proper man. Let me see now:
> To get his place, and to plume up my will
> In double knavery. How, how? Let's see:
> After some time, to abuse Othello's ear,
> That he is too familiar with his wife.
> He has a person and a smooth dispose
> To be suspected – fram'd to make women false.
> The Moor is of a free and honest nature,
> That thinks men honest that but seem to be so;
> And will as tenderly be led by th'nose
> As asses are.

I ha't – it is engender'd. Hell and night
Must bring this monstrous birth to the world's light.
(I. iii. 377–98)

Do we discover what the link is? Some phrases in the speech seem to help us. Iago confirms that his intention is to secure the lieutenantship: he wants 'To get his [i.e. Cassio's] place' at line 387. Then, two lines further into the speech, he provides more detail about how he will vex Othello's joy by involving Cassio. 'How, how?' he asks himself, will he 'get Cassio's place'? 'Let's see./After some time, to abuse Othello's ear,/That he is too familiar with his wife.' So, for the first time, a connection is made between a motive (envy of Cassio) and an intention (to make Othello jealous by lying to him about Cassio and Desdemona). But notice how in order to arrive at this interpretation I have had to extrapolate a motive from a combination of what I have described as a feeling (in fact an obsession) in the earlier scene, and what I am describing as an intention ('to get his place') in this one. In fact 'to get his place' sounds like Iago's ultimate intention, in the sense that he is not attempting it in order to do anything else. *How* he will get Cassio's place is *by* abusing Othello's ear. But by this time we have heard a great deal from Iago about his hatred of Othello and his disgust at the Moor's physical love for Desdemona. So we can't be entirely happy with this rather utilitarian explanation of Iago's behaviour. The deflection of his interest, from Cassio in the first part of the first scene to Othello and Desdemona in the second part, is still lodged in our minds – not only because of what Iago says there, but also because of what has happened between I. i. 74 and I. iii. 377 (in which Cassio scarcely figures). So we are inclined to reverse the sequence of Iago's intentions, by reversing the order in which he describes them in the soliloquy. Iago says he wants to get Cassio's place by abusing Othello's ear. But we believe he wants to abuse Othello's ear by getting Cassio's place, through making Othello suspicious of Cassio and Desdemona.

In life and art, intentions are often multiple and single at the same time. Here, Iago wants to get Cassio's place and abuse Othello's ear. Whether or not we believe Iago about which intention logically and psychologically precedes which, it must be the case that one of them is governed by the other.[10] One of these intentions would be identified in an answer to the question: what does Iago want to achieve? That was why I described it as an ultimate intention – ultimate in the sense that, having been realized, there is no other intention (in the 'sequence' of intentions) left to be realized. The other intention would be an intention to be identified in an answer to the question: how does he propose to achieve it? But the 'how'-intentions are all included in the 'what'-intention. If Iago ultimately intends to get Cassio's place, then the intention about vexing Othello's joy or abusing his ear is subordinate to that intention and included in it. Every action comprising the business of vexing or abusing is an intentional act which can ultimately be explained only by reference to the governing intentional act of getting Cassio's place.

A lot of what I have to say about the motive – intention – act continuum in *Othello* depends on getting this point clear, so I shall direct the reader to a philosophical treatment of it.

In her book about *Intention*,[11] G. E. M. Anscombe invents a little parable about a man who is trying to eliminate a fascist cell by contaminating the source of its water supply with a deadly poison. She goes into great detail about the variety of ways one might go about describing this man's actions. I shall simplify by reducing these to four, i.e. four descriptions of a single intentional act which I have described as the ultimate intentional action. These are: (a) the man moves his hand up and down; (b) this operates a pump; (c) this replenishes the water supply; (d) this poisons the fascists. No doubt there are any number of other descriptions of intentional acts which might be inserted between (a) and (d). Anscombe offers the example of forcing the water to run along the pipes. But the main point she wants to make is that the fourth description 'swallows up' (her phrase) all the earlier descriptions in the series:

> When we speak of four intentions, we are speaking of the character of being intentional that belongs to the act in each of the four descriptions; but when we speak of one intention, we are speaking of intention *with which*; the last term we give in such a series gives the intention *with* which the act in each of its other descriptions was done, and this intention so to speak swallows up all the preceding intentions *with* which earlier members of the series were done. The mark of this 'swallowing up' is that it is not wrong to give D [i.e. my (d): Anscombe uses the letters A, B, C, D] as the answer to the question 'Why?' about A; A's being done

with B as intention does not mean that D is only indirectly the intention of A... If D is given as the answer to the question 'Why?' about A, B and C can make an appearance in answer to a question 'How?'. When terms are related in this fashion, they constitute a series of means, the last term of which is, just by being given as the last, so far treated as an end.[12]

Let us see what happens when we apply this parable to *Othello*. The equivalent of the man's poisoning the fascists is Iago's getting Cassio's place (i.e. that is the way he puts it to himself in the soliloquy). That is D, or (d), or the 'swallowing up', or the ultimate intention. Iago asks Anscombe's own question: 'How?', and answers it by saying that 'After some time [he will] abuse Othello's ear,/That he is too familiar with his wife'. It is crucial here to notice that Iago interprets the intention to abuse Othello's ear as a description of the same thing that was described earlier as getting Cassio's place. In doing so, he assigns it to a different category from, for example, persuading Emilia to steal Desdemona's handkerchief. This tells us something about how Iago realizes his ultimate intention, but it is not another description of it. Stealing Desdemona's handkerchief (in the event just finding it dropped on the floor) doesn't automatically lead to Iago's getting Cassio's place, but abusing Othello's ear does automatically lead to it. Abusing Othello's ear entails getting Cassio's place in the same way as the man's moving his hand up and down entails poisoning the fascists. Of course, it might not actually happen. The pipe might get blocked or the fascists might get suspicious. But that wouldn't stop the man's moving his hand up and down being a description of an intentional act which belonged to a sequence of intentional acts which ended with the fascists being poisoned – i.e. it wouldn't stop it being that, even if the fascists weren't poisoned. In the same way Iago might not achieve his ultimate intention of getting Cassio's place. Othello might kill Desdemona instead of Cassio, or Iago's design might be suspected after the damage had been done. This is in fact what happens in *Othello*. However, our interpretation might be different if he gets his priorities the wrong way round in the soliloquy, as I have suggested he does. What if poisoning the fascists turned out to be a 'how'-intention that was ultimately swallowed up in the man's 'why'-intention to move his arm up and down, but the man didn't know this? What difference would this make to our description of intentional action – in this case Iago's intentional action to abuse Othello's ear, which I suppose is ultimately swallowed up by the intention to destroy Othello?

First, is this what happens? Are there sufficient grounds for supposing that Iago has reversed his intentional position in this soliloquy? So far the evidence I have supplied has been taken from what we have discovered about Iago's attitude towards Othello earlier in the play. But there is evidence in the soliloquy too, reinforced by what Iago discloses in his second soliloquy, at the end of II. i. The references to getting Cassio's place and abusing Othello's ear occur towards the end of the first soliloquy, and Iago has much more to say about Othello in the lines preceding them. The first three and a half lines are there merely to ensure that Roderigo is dismissed from our and Iago's thoughts. Clearly his status in the play is not much more important than the handkerchief. Although he is skilfully characterized, his role in the play is almost entirely instrumental. His job is to hurry up Iago's plot against Othello, and then to act as an agent in that plot (but an agent firmly under Iago's control).

Thereafter Iago makes three claims which, taken in the context of his statement about getting Cassio's place as an ultimate intention, are extraordinary.

The first is at line 384: 'I hate the Moor.'

The second follows immediately, and is loosely related to the first claim by a simple coordination: 'And it is thought abroad that 'twixt my sheets/He's done my office.' He is claiming that there is a rumour going about that Othello has had an affair with Emilia. Iago, out of mere suspicion, intends to even with him for it. The word 'even' appears in the second soliloquy – 'nothing can and shall content my soul/Till I am even'd with him wife for wife' – in which the motif of Iago's jealousy over Emilia is much more pronounced: 'For that I do suspect the lustful Moor/Hath leap'd into my seat; the thought whereof/Doth like a poisonous mineral gnaw my inwards' *etc.* This adds another motive for Iago's design on Othello, as does the reference to his love of Desdemona near the beginning of the second soliloquy. This almost completes the list of Iago's specific motives. The only other one appears near the end of the second soliloquy, where he claims that he 'fear[s] Cassio with my night-cap too', i.e. Cassio as well as Othello has slept with Emilia. I shall return to this proliferation of motives later.

In the meantime there is a third claim in this first soliloquy to be explained. It occupies the second half of the line about getting

Cassio's place, and is worded differently in different editions of the play as 'plume up my will' (Folio and second Quarto) and 'make up my will' (first Quarto). In both versions the phrase doesn't mean much more (or less) than that Iago wants to have his way merely for the sake of having it.

There are three further claims Iago makes in the first soliloquy, augmented but not significantly altered by those he makes in the second.[13] He hates Othello; he has at least two more specific motives for wanting to get his own back on him (and one more for wanting to get his own back on Cassio); and he wants to act wilfully just for the sake of acting wilfully.

Some observations on Iago's statement of his motives in this soliloquy. First, regarding their number: the more they proliferate, the less convincing any one of them taken singly seems to be. Two of them are both specific and immediate – Cassio has only recently been awarded the lieutenantship, and Desdemona has only recently married Othello. A third is even more so – Roderigo wants Iago to turn what seems to be a disaster into a triumph: if not he wants his jewels back. The other motives – Cassio and Othello have cuckolded him, and he is in love with Desdemona – appear to be much longer-standing grudges, given the peculiar double time scheme of this play. In any case, the last of these is probably thrown in as a belated mark of respect for the Cinthio source, in which the cause of Iago's plotting unambiguously lies in his passion for Desdemona.

The more one looks into these motives, the more the immediate ones take on the character of opportunities for the satisfaction of the other, deeper-seated ones. In the early scenes, Cassio's appointment looks much like the 'real' reason for Iago's forming whatever intentions he forms. Then Othello's marriage to Desdemona seems to take over from that as the principal object of Iago's attention. At the very least, vexing Othello's mind against Desdemona by making him suspicious of Cassio will kill two birds with one stone. Then, in the soliloquies, the emphasis shifts again. A preoccupation with Othello's marriage to Desdemona seems to lie behind the specifically sexual jealousy that occupies so much space here. We begin to suspect that these latent, hidden motives must be the 'real' ones, just because they are hidden, latent. Or we would do if they weren't so ridiculous. The one about Iago's love of Desdemona is an irrelevant bonus from the source. We know that neither Cassio nor Othello has slept with Emilia, because Emilia is a trustworthy character and she would have told us if they had. In fact she does tell us that Iago's suspicions of Othello are unfounded (IV. ii. 146–8)'; and I don't think anyone could reasonably interpret her lines at IV. iii. 84–101 as an oblique confession that she has slept with Cassio. For one thing she still believes that Iago has used her well, and for Emilia a wife's not having been used well is the only thing that can excuse infidelity to a husband (102). So the further we move through the first two acts of the play, i.e. before Iago begins to put his intentions into practice, the more our suspicions are confirmed that two almost contradictory things are going on. As Iago's intentions grow more dangerous, his reasons for having them become less immediately practical (he wants Cassio's place) and more psychologically deep-seated (he is jealous of Othello). But at the same time as his reasons become more psychologically deep-seated, they become less plausible.

Iago obviously isn't in love with Desdemona. In any case for Iago love is not distinguishable from 'a lust of the blood and a permission of the will' (I. iii. 333–4). He says that, though lustful, he really 'loves' Desdemona only 'to diet his revenge' on Othello. He can't really believe that either Othello or Cassio has slept with Emilia, because he has no grounds for thinking so, and he is not a stupid man. He admits he is using the 'mere suspicion' as a sort of perverse driving force in his plot to undermine Othello. And this is the most significant feature of Iago's articulation of his motives. They are based on fantasies. Iago's sharp intelligence is placed in the service of a circumscribed practical intention, but his uncontrollable fantasy life widens the boundaries and scope of the intention to such an extent that it is transformed.[14] After all, when does Othello actually offer him the lieutenantship? When he has sworn to murder Cassio. And at what stage in the play does he swear to murder Cassio? When he knows that, at his own prompting, Othello has embarked on a course of bloody revenge that can't avoid having the effect of leaving no general for him to be lieutenant to. This is at the end of III. iii, where Iago's mysterious request to let Desdemona live (481) is met by Othello's damning her thrice and promising to 'furnish me with some swift means of death/For the fair devil' (484–5). By now the audience can see where Othello's feeling will lead, if it is not corrected by the exposure of Iago's deception. And Iago must know too. It cannot lead to a permanent appointment. To pretend that Iago's motives and intentions have any sort of practical aim, even as early as I. iii, is to misconstrue the play.

The sheer number of Iago's motives, and the order in which they make their appearance in *Othello*, strongly suggest that no one of them accounts for his intentions or his actions. Vexing Othello's joy is an indeterminate intention that swallows up whatever determinate intentions Iago discloses to us. These include abusing Othello's ear and getting Cassio's place, and I have argued that getting Cassio's place is closer to Anscombe's B than to her C in Iago's series of intentional acts. It is certainly not D, even though at times he pretends it is. It can't be, because D is vexing Othello's joy, or whatever Iago means at the end of the first soliloquy when he says 'It is engendered', where 'It' is defined in no more precise terms than as a 'monstrous birth'. But there is an enormous difference between something as imprecisely defined as 'a monstrous birth' and something as clear and definite as 'poisoning the fascists'. An intention generated by motives as obscure or fantastic as Iago's have proved to be is itself likely to be obscure and fantastic. Above all it is likely to be open-ended. The series of Iago's intentions begins with something that could be described as a specific aim, goes on to include aims of greater and greater indeterminacy, and stops at something he cannot articulate except in mysterious and enigmatic terms. Only the act eventually discloses what it is, and converts suspicion, even prediction, into knowledge. As Iago himself puts it, 'knavery's plain face is never seen till us'd' (II. ii. 306). Remember that in Anscombe's description of intentional acts, the combination of these acts is referred to as a series. Really, though, it is a container in which members of the logical sequence of how-intentions swallow one another up as they move closer and closer to what is eventually discovered to be the why-intention which then becomes the ultimate intentional act. In *Othello*, Iago's ultimate intentional act is 'what happens in V. ii', and that is an act which all of Iago's how-intentions lead to, but which he doesn't know they lead to. For him, V. ii is just 'It', or 'a monstrous birth'. In Iago's case the fantasy motives seem to bypass the ABC intentional acts to work as a direct pressure on the D intentional act. What A, B and C do is serve these obscure motives by expanding the limits of Iago's plot from 'getting Cassio's place' to 'abusing Othello's ear'. Thence they precipitate D which, articulated early on as 'vexing Othello's joy', discloses what it really is, what 'vexing Othello's joy' means, in the events of V. ii. That is what I had in mind when I described the motive – intention – act continuum in *Othello* as a problem for interpretation.

I want to return to the first and third claims in the soliloquy, which I described as extraordinary. The most interesting feature of these claims is not the way Iago articulates them but the way he articulates the connections between them. In both instances this is by simple coordination: 'I hate the Moor;/*And* it is thought...' 'To get his place, *and* to plume up my will.' In the first example a very forceful and absolute claim about feeling is made, to be followed by a very particular account of a state of affairs. It is natural to suppose that the first is explained by the second. But that is not what the syntax encourages us to do. There is one thing, a feeling; and there is another thing, a state of affairs. The syntax doesn't say that the one depends on, or is caused by, the other. And the claim to a feeling is made before the account of a state of affairs. It is the combination of this temporal priority and the loose coordination with what follows that makes the hatred sound so absolute. It exists in relation to knowledge of the state of affairs, but not in dependence on it. It is possible to assume that the hatred precipitates thoughts about a state of affairs which are introduced merely to provide the hatred with a (possibly spurious) reason for being there.

In the second example we are faced with the expression of two intentions also related to each other by a loose coordination. Here the more specific intention 'to get his place' comes first. The more general and objectless one, 'to plume up my will', follows it, but doesn't depend on it. Nor does 'getting his place' depend on 'pluming up my will'. Instead, the two intentional acts are free-floating – 'double knavery' implying that they are to be conceived as superimposed the one upon the other, but in no prescribed order. The one is an intention which lacks substance, the other a motive which lacks a cause. Both of these 'lacks' or 'omissions' have to be made good or accounted for. And this can be done only by producing illogical threads of connection: in the one case back into a dream-land of fantastic grievances; in the other forward into a design of surreal irrelevance. Reality and relevance are not things that provide grounds or motives for actions. They are what actions prompted by fantasy and self-will bring into being. In Iago's case what really and relevantly happens as a result of the indulgence of his fantasy and self-will is peculiarly horrible. That is because his fantasies are what they are, and his self is what it is. But fantasy and wilfulness, operating within the field of motive and intentional action I have tried to describe, don't have to create something as horrible as what happens in *Othello* V. ii. In different circumstances, they might create something more like *Othello* itself. Again, Iago's soliloquy suggests how this might come about.

After Iago's words about his 'suspicion' of Othello and Emilia, he doesn't describe what he is going to do in terms of motive or intention, but in terms of purpose:[15] 'The better shall my purpose work on him.' When Macbeth recognizes he is coming to a decision to assassinate the king, he describes the movement from inclination to decision as an 'intent'. He proposes to himself a definite course of action, and that is his 'intention'. But the drift of my argument about *Othello* is that Iago never proposes any such definite course of action to himself. He talks about putting Othello 'at least into a jealousy so strong/That judgement cannot cure' (II. i. 295–6), and he says that out of Desdemona's goodness he will make 'The net that shall enmesh them all' (II. ii. 350–1). But he never proposes acting with the specific intention signalled by Macbeth's use of words like 'bear the knife' or 'Taking-off' (of Duncan). And he never imagines the precise details of what he is about to do, as Macbeth does when he sees 'gouts of blood' on the dagger of the mind which is as 'palpable' as the dagger – 'this' dagger – 'which now I draw'. One might reply that this is not surprising because it is Othello, not Iago, who murders Desdemona. But this is too simple. It is a fact that vexing Othello's joy and 'practising upon his peace and quiet,/Even to madness' actually does mean what happens in Othello V. ii. as much as the bloody dagger in *Macbeth* means what happens between Acts II. i. and II. ii. So if Iago's mind had been as vividly intentional as Macbeth's was, he would have seen the pillow coming down over Desdemona's face and the blade thrust into Othello's body. In both tragedies, intentions issue in specific actions or in conceptions of action which are closely related to them. Purposes are much less specific. A character can have a purpose without having a specific end in view, but an intention is deliberately and of its nature end-directed (though the issue is clouded by the fact that, in Gustafson's words, 'both plans and purposes are expressed in the forms of intention').[16] I want to apply this distinction to Iago's role in *Othello*, and then to Shakespeare's role in the composition of *Othello* and *Macbeth*. What was Shakespeare's intention and what was his purpose in the act of writing these plays? How does our description of Iago's 'purpose' in *Othello* help us to answer this question?

Far from being what Coleridge described as a 'motiveless malignity',[17] Iago is a character embarrassingly over-endowed with motive. The difficulty is rather that his motives don't account for his intentions, and his intentions can be made to account for his actions only by recourse to the special pleading adopted earlier in this chapter. I am suggesting that we agree with Wittgenstein that intentions are private states, but that expressions of intention are available to public scrutiny. Wimsatt and Beardsley failed to discriminate between intentions and expressions of intention, and therefore conflated the meanings of intention and motive. Anscombe seems to agree with Wimsatt and Beardsley here. She will only go so far as to separate intention from motive by distinguishing elliptical from non-elliptical forms of expression. So, if a man were to be asked to explain what he was trying to achieve in a business transaction he would be most likely to say he wanted to 'gain' something. A gambler would want to 'win' something. These words – 'gain', 'win' – are non-elliptical expressions when used of intention, but elliptical ones when used of motive, where the technically correct answer would be 'desire of gain' or 'desire to win'. In other words the terms are almost interchangeable, and usually they are interchanged. When one is talking about motive and intention, one is looking at the same thing operating in a slightly different psychological time-space. Clearly the desire precedes the hypothetical satisfaction of desire. Intentions are bound to be speculative and practical, motives causal and explanatory.

But are these really the only differences? Don't we use 'motive' in two different senses, one of which *is* very close to intention, the other further removed? Anscombe's motive is a mental act that answers the question 'what for?' with a response that will probably begin 'in order to ...', and then proceed to an explanation of the intended effect. Obviously this is very close to intention. But there is another way of defining motive: as the mental act that answers the question 'why?' – this time with a response that will probably begin 'because...', which leads away from explanations about what something happened *for* (a forward-looking explanation) to what it was that set in motion the something that happened. 'Desire for gain' might still be an answer to a question about motive of the 'why' type, in the form of 'because I had a desire for gain'. But another permissible answer might be 'because I enjoy taking risks', or 'because I was brought up by communists who repressed my interest in capitalism', or even, to come closer to the 'what for?' answer, 'because I was brought up by communists who repressed my desire for gain'. In this last example, the non-elliptical form of the motive-cum-intention word 'gain' is used, but more is added to explain its being there. And although it could be argued that the most immediate and specific motive was thereby proved to have been 'desire for gain', one might counter-argue that immediate motives aren't necessarily more convincing ones, or truer ones, than those that are more distant. Freudian psychoanalysts would claim the opposite, that the inhibited motive is more likely to be the truer one, in the sense that knowledge of it would be more likely than knowledge of the immediate 'intentional' motive to increase the chances of successfully treating the patient.

It is in accordance with this latter model of the inhibited, indirect, or 'why?/because...' motive that I claim Iago's motives don't account for his intentions. It is perfectly reasonable to say: Iago resents Cassio's being preferred over himself in the matter of the lieutenantship, and that is his motive; Iago plots to usurp Cassio's position, and that is his intention. But having said it, how much of *Othello* have you understood? On the other hand it doesn't sound reasonable to say: Iago hates Othello, and that is his motive; Iago plots to usurp Cassio's position, and that is his intention – if you are implying there is some sort of causal connection between motive and intention in each case. But having said this, you might be on the way to understanding something about *Othello*, because the discrepancy between the statement of motive and the statement of intention could open up all sorts of relevant considerations about Iago's state of mind and its effects on his actions. You would have to try to justify the discrepancy, or show that somewhere under the surface of the statement there isn't a discrepancy. And when you have got far enough under the surface to perceive the fantastical versions of Othello, Cassio, Desdemona and Emilia concealed there, you will be in a position to say that from an audience point of view Iago's motives don't account for his intentions, because his motives are fantastic and his intentions appallingly real; but that from an Iago-eye point of view they do account for them, because for Iago his motives are real (after all, he invented them) and his intentions are not so much fantastic as phantasmagorical. They are too closely related to his motives. That is why they take on the colouring, but not the firm, clear outline, of those images of 'carnal stings', 'unbitted lusts' and leaping into his seat that propel him forward into his vague intentions and hideous acts.

MACBETH AND SEARLE

Everything happens by accident and nothing can be predicted.
Dungeons and Dragons Players' Manual.

Writing about Macbeth's use of words like 'vault', 'o'erleap' and 'hors'd' – in his soliloquy at I. vi. – I suggested that Shakespeare's selection of them was affected by 'a larger-scale intention which, when realized, and expanded to its proper limits, became the play'. I should now like to rephrase this suggestion by persisting in describing the selection of the words as an intentional act which at a conscious level of Shakespeare's mind must have been voluntary and deliberate, adding only that the 'larger-scale intention' would be better described as Shakespeare's purpose.[18] Then it becomes possible to argue that his purpose was served by the intentional act of selecting these words, but also that the fact of their having been selected had a modifying effect on his purpose.

The image of o'erleaping in *Macbeth* is a useful one, suggesting as it does a voluntary act carrying the person who performs it beyond the point at which he expected his intention to be realized, towards another point which may or may not be consistent with the further exploitation of his intentional 'drive'. Macbeth's o'erleaping ambition (in the soliloquy) carries him beyond the perceived object of his ambition, and his o'erleaping of the step that represents Malcolm's prior claim to the throne of Scotland (in the aside) carries him beyond the point at which it is legitimate for him to aspire (thus paradoxically making 'I must fall down' and 'or else o'erleap' not genuine alternatives but different descriptions of the same terrible act). In the same way Shakespeare's habit of o'erleaping the original end of his imaginative intentions carries him beyond the limits set by what was earlier deemed to be his governing intentional act. This is what I mean to suggest by using the term 'modification of his purpose' in this context. One of the differences between Macbeth and Shakespeare is that the image of falling (twice repeated in relation to the 'o'erleaping'), in the sense of losing a flexible intentional impetus that opens up rather than forecloses opportunities, applies to Macbeth in his accomplishing a real murder, in a way that it doesn't apply to Shakespeare in accomplishing the description of a murder. This is the case even though Shakespeare so familiarly inhabits Macbeth's mind as he gestates and expresses his intention. After all, Shakespeare provided himself with manifold opportunities for both access to and egress from Macbeth's consciousness. But for Macbeth there is no way out of it.

Anscombe set before us a situation in which the highest-level or ultimate or 'final' intention of the man moving his hand up and down was to poison the fascists. I compared this man with Macbeth somewhere around the point at which he refers explicitly to his intent, or sees a bloody dagger, or tells Lady Macbeth 'I go, and it is done'. Somewhere at or between these three points Macbeth fully forms his intention to kill Duncan. The difference is that after the man has poisoned the fascists he disappears from Anscombe's book. Rightly so, because he has done his job of illustrating the series of intentions which is completed by the description of intentional action D: he has poisoned the fascists. But the equivalent job performed by Macbeth is described in a series of intentional actions beginning with the point at which Macbeth, having formed his intention to kill Duncan, sets in motion an action which will automatically achieve that result. I would locate that point at II. i. 42: Macbeth grasps his own dagger as he announces that the dagger of the mind 'marshall'st me the way that I was going'. It ends, of course, with the actual murder, which occurs immediately after the soliloquy in which these words appear. Having formed the intention of murdering the king, Macbeth performs the murder in an intentional action made up of the series: grasps the dagger, sees the phantasmal dagger, strides to Duncan's chamber, is invited by the bell, (stabs Duncan) (bracketed because we don't see this happening). All of these intentional actions are swallowed up in the overall intentional action of murdering Duncan. But each one of them is a term in the description of how it is done. Shakespeare is in firm control of Macbeth's doing it. Within the space of these 23 lines nothing is going to deflect Macbeth from doing the deed,[19] and nothing is going to deflect Shakespeare from describing his doing it. The intentional acts of Macbeth and Shakespeare coincide with each other line by line: Macbeth murdering the king; Shakespeare describing Macbeth murdering the king.

However, they are only 23 lines. Before them we have observed Macbeth struggling at first to abort and then to realize this intention. Afterwards we shall see him living with the consequences of its having been realized. And we shall become very much aware that among the starkest of its consequences is the fact that, with hindsight, we can see that in a very important sense murdering the king wasn't really D at all. 'To be thus is nothing,/But to be safely thus'

(III. i. 47–8). Now the intention is reinterpreted as Macbeth's becoming king himself, being safely the king, and being all that it means, or should mean, to be the king. It may be true to say that all along the idea of killing the king has been inseparable from becoming king as a necessary consequence of the old king's death. Certainly, in spite of the plot business of Malcolm's and Donalbain's escape from the castle, that is exactly what happens. We could say that E exists, because D is not the ultimate but the penultimate member of the ABC *etc* series. D is killing the king; E is Macbeth becoming king. The first doesn't just lead to the second; it *is* the second, *is* what it was *for*. D automatically entailed E just as A, B and C automatically entailed D. So the addition of an extra member to the intentional series doesn't invalidate the series. It merely lengthens it in a sense that is as much logical as temporal.

The complication arises not out of Macbeth's becoming king, but out of his not becoming the kind of king he wants to be – secure, legitimate, respectable. Those words ought to be an important part of the definition of what it is to be a king – as Duncan was a king before Macbeth murdered him. But can this really have been a part of Macbeth's intention? I don't think so. Nevertheless it has to be confessed that Macbeth was aware of the future complication some time before he formed the intention of bringing about the state of affairs that would inevitably produce it. Almost the whole of the I. vii. soliloquy is given over to speculations, hopes and fears about the 'consequence' of doing what he plans to do. 'If the assassination/ Could trammel up the consequence, and catch/With his surcease success...' But Macbeth knows it cannot. Therefore his intention to murder the king and seize the crown rides roughshod over his knowledge that the real Duncan will not die but live on in the legitimate form of his true heirs, Banquo's issue, and that the kingship he will have achieved will be a kingship in name only, not in substance. Macbeth can know this, but at the same time he cannot make it part of his intention – though inevitably it follows from his intention.

An alternative guide to Anscombe on the philosophical aspects of this complication is John Searle, in his book *Intentionality*.[20] Searle uses the phrase 'accordian effect' to describe a sequence of complex intentions. His example is Gavrilo Princip's murder of the Archduke Ferdinand in Sarajevo. In his account of this, Searle tables a list of events where each member is systematically related to those preceding and succeeding it. Thus in a single intentional action Princip in effect did six things: he pulled the trigger, fired the gun, shot the Archduke, killed the Archduke, struck a blow against Austria, and avenged Serbia. So far, this is no different from Anscombe's list – the one that extended from the man's moving his arm up and down to poisoning the fascists – except that she included fewer members. She might have reached Searle's total by adding to her list 'he struck a blow against Hitler, and avenged the Jews', if avenging the Jews were the man's ultimate intention, i.e. the intention that 'swallowed up' the rest. The number of items on the list is not important. What is important is Searle's point that we can't go on indefinitely extending the number of members of the sequence. We can't do this because, although in his example there are causes and consequences of Princip's intentional action that undoubtedly need to be mentioned if we are to give a full account of it, these causes and consequences are not in themselves parts of that intentional action, they don't belong to the series of intentional acts. For example, before Princip pulled the trigger, he produced neuron firings in his brain and he contracted certain muscles in his arm and hand. While actually doing the firing, and thus killing the Archduke and avenging Serbia, he moved a lot of air molecules. And the result of Princip's action, relating to it as an effect to a cause, was that he ruined Lord Grey's summer season and started the First World War.

The neurological and muscular events are irrelevant to my argument, and the side effects (the equivalent of moving a lot of air molecules) are difficult to disentangle from the consequential ones. It is the consequential events that produce the complications I was referring to in my discussion of *Macbeth*. Searle calls them 'the unintended occurrences that happened as a result of his action', and he excludes them from the list of intentional actions because they don't satisfy the requirement of counting as a true answer to the question 'what are you now doing?', where that question asks 'what intentional action are you now performing or trying to perform?' In the case of Gavrilo Princip's action it is relatively easy to determine which descriptions are descriptions of intentional actions and which are not, because it is always easy to answer a question in the form of the one quoted above and to know whether the answer you have supplied is a true answer or not. Nevertheless as I move down Searle's list (and when I move down the amended list from Anscombe) I am struck by the progressive increase in what one might call the psychological content. Pulling the trigger is just a tiny bit more mechanical and 'definite' than killing the Archduke (in the sense that you would be more certain about the former than the latter if asked what intentional action you were performing). More to the point, you would be saying something much more intelligible and unambiguous if you answered that you were pulling the trigger, or even killing the Archduke, than if you answered that you were avenging Serbia.

I am not trying to detach the last items in the series from the list of genuine intentional actions. Clearly they belong there in a sense in which starting the First World War does not. From the literary critical point of view, drawing parallels with Macbeth's action in killing the king, it is the movement in the direction of psychological description that is interesting. For here we have not only the equivalents of spoiling Lord Grey's summer season (terminating Malcolm's and Donalbain's residence in Scotland, for example), but also of avenging Serbia (becoming king). Gavrilo Princip avenged Serbia, but he may or may not have acquired the emotional satisfaction that is suggested by the use of the word 'revenge'. 'Succeed to the throne' sounds like and is the end of something. But in *Macbeth* it is also *not* the end of something, and it *is* the beginning of something else, variably predicted and not predicted. In the same way 'avenge' sounds like and is the end of something, but in Princip it might well have been also *not* the end of something and it *might* have been the beginning of something else. In each case the something else that began is not just the out-of-series First World War or coronation. It is also the implicit but mysterious psychological content of 'avenge' and 'succeed to the throne' which hovers awkwardly at the edge of the series: difficult to tell whether it is in or out.

Intentional acts – in the sense that Wittgenstein and Anscombe understand that phrase – are relatively simple. And they *are* acts. They are never states of mind, or acts of which it can be predicted that they will have the effect of creating particular states of mind. The man moving his hand up and down will certainly poison the fascists, unless some accidental and external factor supervenes which prevents the poison from being carried along the water supply or forces the fascists out of the house before they can drink the poison. But whether he will derive satisfaction, a sense of futility, hilarious joy or depression from having done so, cannot be predicted with any degree of certainty. One can, of course, hope to be satisfied by intending a course of action. But a state of satisfaction can never be a member of a series of intentional actions. Nor did Macbeth expect it would be in his case. Having agonized over the psychological and spiritual consequences of the realization of his intentional act, he set on one side his suspicion of what these might be, and then acted. What came afterwards, i.e. what Macbeth discovered his spiritual condition to be, might be interpreted as a necessary consequence of the act. But it was not a necessary consequence of Macbeth's intentions precipitated into the act, because by the time they were so precipitated Macbeth had removed his conscious knowledge of them from the self that had submitted to the demands of the decision he had made. This is evident in his dagger soliloquy, where the 'bloody business' 'informs' his eyes, and he describes himself as being at one with 'the present horrors' of the time.

Intentional actions have to be brief if they are to be successfully implemented. And they do have to be actions. As such, they are precise and definite. Where a discrepancy arises between the performance of an intentional action and an effect the action is expected to bring about, this is because there are fewer members in the series of intentional actions than was originally thought to be the case. One of the how-intentions it was assumed were to be swallowed up in the why-intention at the end of the series is itself discovered to be the final intention. This intention is now the one that explains why the other intentional actions were performed. The other 'effect' must therefore exist in a totally different relation to the series of intentional acts it was believed to complete. I have already indicated that *Macbeth* is unusual because of the 'success' of Macbeth's intentional action in murdering Duncan in the second act. But Macbeth isn't just a play about the murder of Duncan. It is also a play about the circumstances in which that murder was carried out, and it is a play about what followed from that murder. What, then, is the connection between Macbeth's fears for the future which preceded the intentional series, and the legitimizing of those fears in the events that followed it?

Referring to Macbeth, we might define it as the confirmation of a psychological prediction. But I am not exclusively referring to Macbeth. I am referring also to what Shakespeare is doing with Macbeth in the play *Macbeth*. In dramatizing Macbeth's intentional act of killing the king, Shakespeare enacts the intention with Macbeth. Some of the words Macbeth is given to speak in the dagger soliloquy make it clear that Shakespeare is doing other things too, things that show him mediating between Macbeth and the audience. Fundamentally, Shakespeare is performing on a linguistic level what Macbeth is performing on a psychological level.[21] They are at one in their intention, differing only in the existential or artistic responsibility they accept for it. This is not true, though, of what surrounds the expressed intention. It is not true of the patterning of events within which the intended act takes place, or the connections between the events, or the emphasis placed on the representation of some events as compared with the emphasis placed on that of others. Macbeth performs many other intentional acts, and when he does, at the moment at which he does them, Shakespeare shares in that performance – on a linguistic level which touches on the psychological/existential level at all points. But Shakespeare is doing a lot of other things too. For example, he is performing Lady Macbeth's intentional acts – on that same expressive or linguistic level. And Banquo's. And Malcolm's. And he is organizing the representation of these acts in meaningful sequences. By doing so, he is creating a narrative, generating and sustaining a rhythm, that binds events together in a way that is related to, but is not the same thing as, a motivated or causally intelligible series.

For Macbeth it took a great deal of fumbling and self-deception and perverse courage to arrive at the position in which he could perform the intentional act of killing the king, during that brief space of dramatic time between the opening of the soliloquy at the end of II. i. and the beginning of II. ii. Having performed it, however, he experiences everything differently. Events can still dismay him. Witness his reaction to Banquo's ghost. But the most characteristic attribute of Macbeth in these middle and late scenes is the enormous gap that opens up between what he ultimately wants but knows he cannot have – a secure and ceremonious kingship – and the intentional acts he performs in order to achieve it. None of these events is in any meaningful sense intentionally related to the goal proposed – as taking up the knife was to dispatching Duncan and becoming king – because now the goal proposed is not so much an act as a state of being. Furthermore the intentional acts themselves contain far fewer members in the series because the gap between preparing to act and acting has closed to such a remarkable degree. Macbeth puts it like this:

> The flighty purpose never is o'ertook
> Unless the deed go with it. From this moment
> The very firstlings of my heart shall be
> The firstlings of my hand. And even now,
> To crown my thoughts with acts, be it thought and done:
> (IV. i. 145–9)

The state of mind these lines express is the one critics refer to when they say that Macbeth's actions become more and more automatic as the play develops. The telescoping of thought and deed, which might have been expected to produce an impression of heroic decisiveness (albeit in an evil cause), in the event produces the impression of a man who is much less heroic than the one who faltered and prevaricated before the murder.

But what was that compared with Hamlet's procrastination before Act V, or Antony's indecisiveness right up to his death at the end of Act IV? Macbeth forms an intention early on in the play. He realizes it in an intentional action that is completed by the end of Act II, after which it is dispersed into a sputtering of brief, efficient but irrelevant intentional acts until he is hacked to death at the end of V. vi. Yet he talks of having a purpose. 'The flighty purpose never is o'ertook/ Unless the deed go with it'; 'This deed I'll do before the purpose cool' (IV. i. 145–53). But he has no purpose in the sense in which I have been using the word in this book. These are mere intentions, the same intentions that flicker uselessly and hurriedly across the stage in so many of the later scenes. Macbeth has carved himself into a position in which any sense of purpose is lost. Unlike Iago, and unlike any of the other great tragic heroes, he has severed the connection of purposiveness between himself and his creator. But Shakespeare's representation of intentional actions, performed by Macbeth and the other characters, has not lost contact with a sense of purpose which remains with him throughout the play. This is because Shakespeare has not identified any one intentional action, or any group of them, with his purpose, or with the kind of purposiveness that is essential to the creation of great art.

The point is vividly illustrated in Soviet critics Eichenbaum and Vygotsky's argument about *Hamlet*. Both of them agreed with Tolstoy's view that Hamlet fails because Shakespeare was unwilling or unable to give the Prince a specific character,[22] but they advanced different reasons for questioning his low opinion of the play on that account. Eichenbaum compares Hamlet's delay in killing the king with Wallenstein's delay in Schiller's drama. Both Hamlet and Wallenstein represent two aspects indispensable for the treatment of tragic forms: a driving force and a delaying force. Instead of a simple movement forward on the path of the subject, or play, we have something like a dance with complex movements.[23] Vygotsky takes up the idea of a driving and delaying force, but distributes them between the hero and the playwright in a manner that raises interesting questions about authorial intention. 'Any artistic method or device', he writes, 'can be grasped much more easily from its teleological trend (the psychological function it performs) than from its causal motivation, which may explain a literary fact but never an aesthetic one.'[24] Accordingly we should ask why Shakespeare makes Hamlet delay, as well as why Hamlet 'in fact' delays. And when we do this we notice, according to Vygotsky, how plot and character are deliberately made to run along different tracks, until they accidentally come together in Act V when Hamlet kills the king. 'Because at all times [Shakespeare] lets us feel, and be aware of the straight line which the action [of Hamlet's killing Claudius] should follow, we are even more keenly conscious of the digressions and loops it describes in actual fact.' These speculations arise out of a significant contrast Vygotsky makes between the handling of character and plot in *Hamlet* and in *Macbeth*, and it raises the question of how far Shakespeare evaded on an artistic level the danger in which he placed Macbeth on an existential one. How did he represent, in a literary form, what Iago (corruptly) and Hamlet (heroically) performed in a series of intentional actions? How, in other words, did Shakespeare free his many intentions from the constraints of one big intention that would have given a shape – but only one, immobile shape – to them all? How did he produce an impression of purposiveness without having any particular purpose?

THE GHOST IN *HAMLET*

> ... it was therefore determined to make this trial of the ... veracity of the supposed spirit
>
> Johnson on the Cock Lane ghost.

Our concern is with Shakespeare's intentions and purposes, only secondarily with those of his characters. It is in the nature of the exercise, though, and of the approach to great literature I am recommending here, that we cannot entirely detach the one from the other.

I have shown how in *Macbeth* Shakespeare collaborates with his hero by running his artistic intentional action in parallel with Macbeth's performative intentional act: representing the act of killing the king in the one case, actually killing him in the other.

I have shown how in *Othello* Shakespeare collaborates with Iago in the same way, noting, though, that here the similarity exists within a basic difference. This is that Iago's intentional acts are not directed to a definite goal or ultimate intention. Of course there has to be a D-intention for all of Iago's A's, B's and C's, but it keeps disguising itself as one or other of the inferior members of the series. We think Iago wants Cassio's place and accept this as his ultimate intention. But no sooner have we done so than we discover it has been replaced (as his ultimate intention) by getting his own back on Othello for sleeping with Emilia, or separating Othello and Desdemona because he is himself in love with Desdemona. Or are these all parallel intentions, of equal weight and value, together subserving the ultimate intention of vexing Othello's joy? If so, what *is* vexing Othello's joy? We find out only at the end of the play, *and so does Iago*. He doesn't really know what his governing intentional action will be until it has been performed. In Macbeth's case this was not so. He knew what the intentional act was, and he predicted what the accompanying spiritual consequences would be. But not both at the same time. That was why the enunciation of his 'intent' came after the havering about 'consequences' in the I. vii soliloquy. Another way of describing the difference would be to say that Iago's activities were purposive within the limits set by his attitudes to Othello, Desdemona and Cassio, but that he had no definite purpose. The purpose grew out of the purposiveness which was the expression of his attitudes to these and other characters, and it became the more clearly defined as his intentional actions created more productive opportunities for it to work with.

When Shakespeare collaborates with Macbeth, he works against his own artistic inclinations – which are to expand, not contract, the number of opportunities for intentional action among his dramatis personae. Luckily, though, Macbeth is not *Macbeth*, and in the play as a whole Shakespeare creates many more opportunities for these actions to be performed. Though Macbeth drives himself into a position in which opportunities for creative action are prematurely foreclosed, Shakespeare does not do likewise. He can make more of Macbeth than Macbeth can make of himself, because the realization of a multitude of other artistic intentions in the play creates a context for interesting himself in Macbeth of which Macbeth himself can have no knowledge or experience.

It is to the creation of such a context, the production of a space in which many intentional actions can be displayed that are not subordinated to any one finally governing intention, that I now want to turn in my approach to the first act of *Hamlet*. I shall concentrate on the ghost, which Dover Wilson described as 'the linchpin of *Hamlet*: remove it and the play falls to pieces'.[25] There can be no clearer example of the expression of an intention in the whole of Shakespeare than there is in his manipulation of the ghost in the first scene of this play. But what intention is being expressed? What intentional action is the ghost performing on Shakespeare's behalf?

It has often been noted that, according to Hamlet, there ought not to be a ghost in *Hamlet* at all. For it is Hamlet who, later in the play (III. i. 79–80), describes death as 'The undiscover'd country from whose bourn/No traveller returns', and this has often been interpreted as meaning either that Shakespeare has been busy constructing another of his infamous Bohemian sea coasts (the third-act hand not knowing what the first-act was doing) or that Hamlet has lost confidence in the ghost and no longer believes it really is one. This kind of inconsequence is not unusual in sixteenth-century drama. There is a close parallel in Heywood's version of Seneca's *Troas*. Here, immediately after Achilles' ghost has appeared to ask the Greeks for the sacrifice of Polyxena, the chorus expresses the view that 'never may a man return to sight/That once hath felt the stroke of Parcas might', and that 'nothing taryeth after dying day'.[26] Yet, as in *Hamlet*, the ghost certainly was there, or something that looked like a ghost was there. As with most other Elizabethan dramatic ghosts, it has an external existence. The ghost of Old Hamlet has a more pronounced external existence than most, having been seen by Marcellus, Barnardo and Horatio as well as by Hamlet. There remains the question of what it is that has an external existence. There is some doubt about that, even among those who are witnesses to it on the battlements.

Marcellus and Barnardo think that the 'majestic' visitor really is the ghost of Hamlet's father, and that his coming to Elsinore portends some political and military development relating to the Norwegian issue. Since the ghost appears in arms, and Old Hamlet has acquired a formidable military reputation by '[smiting] the sledded polacks on the ice', and since we are to learn in the second scene that the Norwegians represent a real and pressing military threat, this seems to be an appropriate response to the ghost, spoken as it is by two soldiers. The only trouble with it is that a sixteenth-century Englishman (or Dane) would have been able to provide no satisfactory explanation of where it had come from. When the ghost speaks to Hamlet at I. v. he describes his present circumstances in convincingly purgatorial terms. But purgatory had been abolished in England during the Protestant Reformation, so perhaps the ghost is lying; or he has come from hell, in which case he is lying even more; or Shakespeare is ignoring the Protestant beliefs of his audience; or he is deliberately taunting them; or an Elizabethan audience would have believed in purgatory anyway, whatever the Anglican church said about it (see *News Out of Purgatory*, anon., 1588); or Shakespeare didn't expect his audience to enquire at all closely into the theological status of the ghost's provenance; or.... Perhaps one should be as non-committal about it as Byron was about the ghost of the Black Friar in *Don Juan* (XVI): 'Once, twice, thrice passed, repassed the thing of air/Or earth beneath or heaven or t'other place.' In any event, the fact that two Danish/English soldiers react to the ghost as they do allows the audience a fair degree of latitude in its interpretation of what the ghost says and does. The same applies to Horatio. At first he thinks the ghost is a 'fantasy' of the soldiers. After he has seen it, he doesn't exactly say what he thinks its status is. He agrees it looks like the king, says it harrows him with fear and wonder, and reacts to it as if it were some kind of demon. (At I. iv. he fears it might assume 'some other horrible form' and draw Hamlet into madness.) On the whole Horatio's opinion in Act I is that the ghost is a demon. At first he adopts the sceptical view that the ghost is either an illusion or a trick. Then he adopts James VI's view that it is a devil in the disguise of a dead man. As with the soldiers, there are arguments pro and contra Horatio – whichever of these views are judged to be uppermost in his mind. Again, the critical point is that Shakespeare's presentation of Horatio's response allows the audience to adopt one or more of several possible interpretations of the ghost's identity and purpose.

Hamlet's assessment of the ghost is even more complicated and uncertain than Horatio's. He finds out much more from it, because it speaks to him. But the more he finds out, the more, not less, ambiguous the ghost's status becomes. The possibilities range across the scale, from 'spirit of health' (or 'heav'n') to 'goblin damn'd'. The spirit Hamlet has seen 'May be a devil' (II. ii. 595) or 'It is an honest ghost' (I. v. 138). Throughout the first three acts Hamlet moves between the Barnardan and Horatian interpretations of the ghost. It could be everything it says it is, or it could be everything it implies it is not – Hamlet's father, or a devilish spirit. It could be an honest ghost and telling a truth for a good purpose, or it could be an honest ghost and telling the truth for a wicked purpose (an argument involving one's interpretation of Elizabethan views about revenge).[27] Or it could be a devil and telling lies, or it could be a devil and telling the truth ('And oftentimes to win us to our harm/ The instruments of darkness tell us truths,/... to betray 's/In deepest consequence' – *Macbeth*, I. iii. 123–6). The permutations are legion. But are they useful? Do they merely confuse us? Or do they open up beguiling possibilities relating to Shakespeare's specific intentions and, through them, to his wider purpose, i.e. the impression of purposefulness which is such a remarkable feature of this play?

In a recent discussion on the role of the ghost in Hamlet, Richard Levin insists there can be no doubt it is what the play says it is – i.e. the ghost of Hamlet's father – and that the more tentative interpretations are the issue of misguided and too simple historical readings of the play. These derive from two basic errors in the critics' line of argument. The first is that 'ideas of the time about ghosts were limited to what the authorities stated in their treatises'. This is certainly an error, but the correction of the error is unlikely to deliver early and conclusive proof of the ghost's identity. The second is that we can induce the author's intentions only from the play itself. This may or may not be true. But if it is true, the fact that 'Shakespeare has ... twice voiced the demonic hypothesis within the action, and twice refuted it' doesn't mean that he has 'very carefully established the Ghost's identity'. It means that the issue has twice been raised, by different characters (Horatio and Hamlet), and twice been refuted, by the same character's (Hamlet's) interpretation of the consequence of its having been raised. The second time it is refuted – very nearly conclusively, I agree – is at the end of III. ii, more than half way through the play.[28]

Although Hamlet seems to be uncertain about what the figure on the battlements really is, and he is supported in his uncertainty by Horatio and the soldiers, the audience feels almost from the start that it is an honest ghost. This is because of what it actually says. When it speaks at first, at I. v. 10–22, it does so in a conventional Senecan manner that sounds rather like the First Player's Pyrrhus speech at II. ii. 446ff. Later, though, the narrative becomes more vivid and less melodramatic:

> Ay, that incestuous, that adulterate beast,
> With witchcraft of his wits, with traitorous gifts –
> O wicked wit and gifts that have the power
> So to seduce! – won to his shameful lust
> The will of my most seeming virtuous queen;
> O Hamlet, what a falling off was there,
> From me, whose love was of that dignity
> That it went hand in hand even with the vow
> I made to her in marriage; and to decline
> Upon a wretch whose honest gifts were poor
> To those of mine!
> But virtue, as it never will be moved,
> Though lewdness court it in a shape of heaven,
> So lust, though to a radiant angel link'd,
> Will sate itself in a celestial bed
> And prey on garbage.

The lines about Old Hamlet's attachment to his beloved wife are dignified and musical; those about his horror at the guilty queen are full of emotional power and what sounds like bitter sincerity. These last (53–7) are not unlike Hamlet's own words to Gertrude at III. iv. of the sincerity of which there can be no doubt. Indeed the whole of this long and circumstantial speech of the old king (41–91), with its detailed explanation of how he died as well as its expression of deep feeling, is utterly convincing about what happened before the play began. Hamlet's immediate inclination is to believe it. Only when the ghost reappears as a voice from the cellarage, and Hamlet behaves towards it as if he were a conjuror and it a devil, does the possibility of doubt arise in our minds. And it is not until the end of II. ii. that Hamlet himself expresses any real, honest doubt – though unexpressed doubt might go some way towards explaining his behaviour at a much earlier stage. But I don't think the audience doubts it at all. Long before we hear Claudius's prayer at III. iii. 36ff., we are aware of his guilt and, therefore, of the ghost's honesty. Or rather, not 'therefore', because sometimes the instruments of darkness tell us truths. We know it is honest because of what we hear it say. But Hamlet hears what it says too, so why is he less convinced by it than we are?

A large part of the explanation must be that Hamlet doesn't want to believe the ghost. Like ourselves, he has the excuse that there are plenty of alternative explanations. They have been dramatized in the reactions of the characters who saw it, and underlined by some of the contradictory things the ghost says. We have that excuse. We have another reason for wanting to disbelieve the ghost, that springs from the close relationship we enjoy with Hamlet himself. We feel the ghost is honest. And we want to identify with Hamlet, who sometimes does and sometimes doesn't feels the ghost is honest. This sets up a tension in our response to the play which is not resolved until III. iv. 103. Here the ghost reappears in Gertrude's closet and Hamlet indicates that he shares our belief in it (it is 'gracious') at the very moment the ghost seeks to deflect his violent anger from Gertrude to Claudius. Hamlet's and our own interpretations of the ghost coincide at precisely the point where the ghost displays for a second time those feelings about the queen that induced us to believe in him back at I. v. This closes a phase during which the discrepancy between Hamlet's and our perceptions of the ghost have enormously complicated our emotional response to the play – a discrepancy made all the more agonizing by virtue of the fact that a part of Hamlet's attitude, or belief, is one that we share. But it is the *whole* of our belief. Only the other, contradictory, half of Hamlet's attitude to the ghost prevents us from making the uncomplicated identification with the hero which his charismatic personality seems to be demanding of us. Vygotsky develops this idea of the audience's double view of Hamlet, depending on whether they are concentrating on the role he plays in the revenge drama or on his speculations about that role and what it tells him (and us) about his character: 'The author actually builds his tragedy on two planes: on the one hand he sees everything with Hamlet's eyes; but then he also views Hamlet with his own – Shakespeare's – eyes, so that the spectator becomes at the same time Hamlet and his contemplator.'[29]

Hamlet tries to dismiss the authority of the ghost by appealing against its ambiguous status, allowing his attitude to be corroborated by what the audience perceives to be contradictory signs in its speech and appearance. As we have seen, this is further emphasized by the different opinions of the dramatis personae who witness it. And he allows the less authoritative version, the stage device, to take its place among the other theatrical impostures he refers to throughout the play (see especially II. ii. 299–307: the speech about sterile promontories and canopies of air). By doing both of these things he succeeds in prolonging his life inside the drama but outside the play. He defends his role in the drama as one which propels him outside the play world into the expressive privacy of the soliloquies, a privacy he shares with us but none of the characters on stage, not even Horatio. It is to facilitate this complex dual response, from Hamlet to the audience and from the audience to Hamlet, that the ghost is thrust into the play and given such an ambivalent status.

The audience's reactions to what is happening in I. i. are rendered even more uncertain than those of the characters on the stage. They must be presumed to know things which, for the moment, we do not know, and this has the effect of creating a looser context, offering larger and more irresponsible interpretative opportunities for us than for Horatio and the soldiers. For one thing, they know more about one another than we know of them. Horatio has had dealings with Barnardo and Marcellus on at least one previous occasion, when they arranged to meet on the battlements. But we are meeting all these characters for the first time. We haven't even heard of them before. In *Hamlet* this turns out not to matter much, because the soldiers are honest and Horatio is the most trustworthy and the most trusted character in the play. Also, they know almost as little as we do about important events that lie in the background of the story and that will affect their and our judgement of what is about to happen. But this is not always the case in Shakespeare. Think of the opening of *Othello*, or of *Antony and Cleopatra*, to see how long it can take us to make up our minds about the characters who appear in them.

In Act I of *Hamlet*, however, the characters soon establish themselves as curious witnesses to an event, not interested parties who are trying to bring an event into being for reasons which might appear inscrutable or obscure. It is the event itself, not the characters' behaviour towards it, that is mysterious. Once again the ghost is the cause of most of the difficulties. I have already enumerated the different interpretations of it provided by the soldiers and Horatio. But what of the way they speak about it, rather than the substance of what they say? In talking to one another they are emptying out the contents of their minds. They don't notice the way the contents are emptied out. But we do, because, unlike them, we are spectators of and participants in the drama. We hear what the words say, and we hear the words being said. For instance, when the formalities of the change in the watch have been completed at I. i. 20, we notice – as Barnardo and Marcellus have no reason to do and make no sign of having done – that the first reference anyone makes to the ghost is Horatio's 'What, has this thing appear'd again to-night?' By having Horatio call it 'this thing', Shakespeare starts off with the most neutral description imaginable. Of course it is perfectly natural that Horatio should use this phrase. But when one places it in the sequence of phrases used about the ghost by all of the characters in I. i. it turns out to have an expressive as well as a mimetic appropriateness. From being 'this thing' it becomes 'this dreaded sight' at 25, and 'this apparition' at 28. Only the stage direction describes it as a ghost, and this is not Shakespeare's. In other words the object to which the characters refer takes some time to achieve any sort of form or identity, and the most precise identity attributed to it before it arrives on the stage is as an 'apparition' – which could mean a ghost, or an evil spirit, or simply something that has appeared. It is not for the moment being pressed into the service of any one definite intention, however generously endowed with dramatic possibilities. It is just something strange which frightens and bewilders those who have encountered it.

When the ghost first appears, at line 40, Barnardo says it comes 'In the same figure like the King that's dead', and 'figure' here is as unspecific as 'apparition' was earlier. In appearance the ghost is a 'fair and warlike form', clad in the armour Old Hamlet wore when he defeated Old Fortinbras in the contest on the ice, moving 'With martial stalk', and generally leaving the impression of warrior-like majesty. Barnardo says three times that it is 'like the king' and obviously thinks it must be the king. But the more sceptical and reflective Horatio refers to it as 'our last king's...image' (80–1), carefully avoiding identifying it with the last king himself, even in the form of his ghost. Before its disappearance at cock crow, he calls it an 'illusion' (127), and afterwards 'a guilty thing', reverting to the neutral substantive already used at the beginning of the scene, but now qualified by the value-word 'guilty'. Even so, it is only 'like' a guilty thing. And Horatio's last reference to it here, as a 'spirit' (171), returns it to the neutral context it has occupied during most of the scene.

On one other occasion in this scene the ghost is referred to as a 'figure', this time a 'portentous figure' – at line 109. But what is it a portent of? The emphasis falls heavily on the military events that are portending. The ghost is clad in armour which is unambiguously associated with the struggle against Norway. At the time the ghost appears, Elsinore is in the throes of military preparation – casting bronze cannon, trading implements of war, building warships. The import of Marcellus's speech, and of the two long speeches by Horatio that follow it, is that the ghost has come to intervene in affairs of state that have to do with Fortinbras's manoeuvres. With the benefit of hindsight from the end of the first half of scene ii we would feel this even more strongly, since by that time we have seen Claudius in conversation with ambassadors he is dispatching to Norway. It begins to look as if Shakespeare's intention in writing the first scene of *Hamlet* was to prepare the audience for a political play (like *Henry V* and *Julius Caesar*). It looks as if the subject will have something to do with the struggle between Denmark and Norway for the territory Norway lost to Old Hamlet, who happens to have died some time before the play begins. What is strange about this is how little it concerns Young Hamlet, who is named only six lines from the end of the scene. When he is named, by Horatio, it is only in the context of what the ghost might be persuaded to say – presumably about the political issue, since no other issue has been mentioned so far.

All the same, I don't think the audience responds to scene i quite as straightforwardly as this. For one thing, the hint about a military intervention doesn't explain why the 'apparition' isn't accepted by Horatio as Old Hamlet's ghost. It doesn't explain the double impression it leaves behind of being both a 'majestical' presence and a 'guilty thing'. Above all, it doesn't explain the religious overtones and undertones which even at this early stage can be heard in the speeches about it – not only Horatio's lyrical duet with Marcellus about the cock crow at 'our Saviour's birth', but also the references to heaven ('by heaven I charge you speak'; 'Have heaven and earth together demonstrated'). When the ghost itself speaks, in scene v, these references will become much more abundant and more emphatic. But even here, before we have an inkling about the personal and spiritual issues involved in the coming of the ghost, the way is being prepared for the absorption of those issues into the structure of the play.

The transition from ghost as political portent to ghost as supplicant for revenge is a long time happening. It is not complete even as late in the play as Hamlet's reference to Claudius's 'Popp[ing] in between th'election and my hopes' at V. ii. 65. Nevertheless, as Act I gets under way, the emphasis does shift from the politics to Hamlet's personal dilemma. But as the shift takes place, the ambiguities take some time to resolve themselves. Hamlet is as curious about the ghost's being armed as he is about any other aspect of it – partly because of the hint of a military purpose, partly also because the coat of armour prolongs the suspense about the ghost's identity. 'Arm'd, say you?' Hamlet asks', 'From top to toe?' 'My lord, from head to foot.' 'Then saw you not his face?' In spite of Horatio's reassurance that he did, these lines show that Hamlet, like Horatio, is uncertain whether the ghost really is his father. By the end of scene ii, he is contradicting himself every bit as much as Horatio and the soldiers did in scene i. 'If it assume my noble father's person,/I'll speak to it,' he promises at lines 244–5. But half a minute later he seems convinced it is 'My father's spirit in arms!'

The soldiers' uncertainty is passed on to Horatio, and Horatio's is passed on to Hamlet, and there is no reason to suppose that in terms of intentional activity Shakespeare doesn't share the uncertainty. There is no evidence he has decided at this stage what the real status of the ghost is. He has not settled the identity of his why-intention by multiplying the how-intentions so assiduously. And in I. v. when the ghost's purpose becomes clear, the question about its identity is still not fully resolved – certainly not for Hamlet. So the purpose is still open to question and the dramatic utility of the ghost is still variable. In many ways the swallowing up of an intentional action tends to create opportunities for the invention of many others, which are emphatically not swallowed up as the dramatic possibilities of the play are expanded.

The coincidence of Shakespeare's and Hamlet's actions also creates dramatic opportunities on occasions when those actions might be described as intentional and unintentional at the same time. It is possible to argue that some actions are intentional under one description and unintentional under another.[30] Anscombe deals with this matter in section 6 of *Intention*. She has arrived at the proposition that 'Intentional actions are ones to which a certain sense of the question "Why" has application.' That certain sense has to do with giving reasons for acting in a particular way. To the question 'why did you start so violently?' there may be no answer other than 'I just did.' So 'why?', there, doesn't mean 'why?' in the sense of 'what is your reason for?' That sense of the question 'why?'

has no application in this instance. On the other hand, to the question 'why did you exclude so and so from your will?' or 'why did you send for a taxi?' (both examples taken from Anscombe) there may be any number of answers which give reasons, and therefore establish as an intentional action what is going on in each question. Starting violently was not an intentional action because whoever was doing it was not aware he was doing it. But there are actions, most actions in fact, that a man might know he was doing under one description (as he knew what he was doing when he sent for a taxi) but not know he was doing under another. For example, a man knows he is sending for a taxi, but he might not know that he is sending for one of H. Jones's taxis or that there is interference on the line.

Anscombe gives the example of a man sawing a plank. In almost every conceivable circumstance one can be sure he knows he is doing that. So it is an intentional action. But one might describe the man sawing the plank in other ways than, simply, 'he is sawing a plank'. One might say he is 'sawing oak', 'sawing one of Mr Smith's planks', 'making a squeaking noise with a saw', or 'making a great deal of sawdust'. Each of these descriptions applies to exactly the same activity as 'sawing a plank', but maybe none of them would be offered by the man as a reason for what he was doing. Asked what he was doing, he might not include the fact that what he was sawing was oak. Therefore sawing oak is not one of the man's intentional actions – though sawing a plank, which is the same thing under a different description, is. Anthony Kenny makes a similar point in a paper on 'Language and mind':

> When we do A by doing B, it may be that we know very well that we are doing A, but do not know without reflection that we are doing B; when we return a serve at tennis there are many movements of hand and arm that we are not normally aware of by which we make the return. Similarly, in performing intellectual tasks – including the comparatively modest one of pronouncing a word or constituting a sentence – there are many sub-tasks we perform without conscious advertence. When we ask what rules or principles we apply in performing these tasks, we are asking what sub-abilities we are exercising when we exercise the ability to use language.[31]

Searle brings us closer to a literary application in his comments on 'unintentional actions'. 'What do people mean', he asks, 'when they say that an action can be intentional under one description but not intentional under another?' He takes Oedipus's marriage to Jocasta as an example. Here we have an action which was intentional under the description 'marrying Jocasta', but not intentional under the description 'marrying his mother'. Searle explains how the different descriptions can apply to the same action,[32] by showing that an intentional action consists of two things: an Intentional component, and an event which is its (the intentional action's) Intentional object. The intention in action (what I have been calling the intention in doing) is the same thing as the Intentional component, but it will not be realized unless it presents the Intentional object as its 'conditions of satisfaction'. The notion of 'conditions of satisfaction' is crucial to Searle's argument about unintentional actions. In the case of Oedipus, the 'total action' – of marrying Jocasta/marrying the Queen of Thebes/marrying (anyone under any description which would be a true description of the person whom he married) – contains some elements that are parts of the conditions of satisfaction of the intention in action and others that are not part of such conditions. In the story of Oedipus, as we find it in Sophocles and elsewhere, the fact that both elements are present in the bare bones of the myth (no version of the Oedipus story known to me fails to identify Oedipus's wife with Oedipus's mother) automatically creates opportunities for the drama which, elsewhere, have to be created from outside the myth, or story. In other words it isn't usually the case that in the stories writers use, unintentional and intentional actions are realized in the same event but under different descriptions, but writers like Shakespeare are doing this all the time, and the way they do it in large part accounts for their success. This, however, is something slightly different from the kind of relationship between the how- and why-intentions I have been discussing in *Hamlet*.

Now let us apply Searle's notion of actions that are intentional under some descriptions but unintentional under others to Anscombe's example of a man sawing a plank. (I favour this example because, unlike Oedipus' marrying Jocasta, it is an action one can imagine occurring uncomplicatedly during a fixed and not especially lengthy period of time.) When Anscombe asked the man what he was doing, she must have done so at some time during the period in which he was doing the sawing. At that time he was sawing the plank, and the reason he was sawing the plank was that he wanted to do it and was conscious of doing it. In Searle's terminology, 'sawing the plank' was both the Intentional component and the Intentional object of the man's activity. The reason the man could not be said to be engaged in the intentional action of 'sawing oak' was that, in our example, the fact that what he was sawing happened to be oak either didn't cross his mind, or it did cross his mind but it didn't seem to matter: ash would have done just as well. Therefore, 'sawing the plank' was a true description of his intentional action, but 'sawing oak' was not, because 'sawing oak' was not an Intentional object which was presented as one of the conditions of satisfaction of the Intentional component 'sawing the plank'. Suppose, though, we imagine the man's sawing the plank occurring over a period of time it was likely to take him to do it. Probably at any stage in the process his intentional action would remain 'sawing the plank'. But it is not inconceivable that half-way through, say, he notices a chalk-mark on the wood and this makes him say to himself: 'Oh yes, this is a plank from Mr Smith's timber-yard', and being reminded of this fact he is put in mind of the occasion when he bought the plank from Mr Smith and got it at a discount because it had a knot in it which would make it harder to saw. Now he remembers the knot, sees it in the wood beneath his saw, and realizes that's why the sawing is taking longer than he hoped it would. So he stops being engaged in the intentional action of just 'sawing a plank' and becomes engaged in the intentional action of 'sawing one of Mr Smith's planks'. Along similar lines we might imagine circumstances that would convert a man's intentional action into that of 'making a squeaking noise with a saw' (getting his own back on noisy neighbours) and all the other activities that become Intentional objects of the Intentional component only when they are presented as being conditions of satisfaction of that Intentional component.

Another way of explaining this substitution of an intended for an unintended component of an intentional action is through Raymond Tallis's claim that the intention will always be framed at a higher level than the action itself, and that therefore the action that realizes the intention will exceed the intention in specificity. 'Intentions, even for simple actions, are, as it were, riddled with unsaturated variables.... Our most deliberate behaviour is porous to the unscheduled.'[33] Before Anscombe's carpenter remembered the provenance of the (oak) planks in Mr Smith's timber-yard, the facts of that provenance and of that material were just such examples of Tallis's unsaturated variables. Then the knot in the wood jogged his memory, and the unscheduled particulars of the work he was engaged upon rushed into his mind and became real features of the intentional act he continued to perform.

The same thing happens on a more imaginative level in Shakespeare. A comparison can be drawn between the man sawing the plank and Shakespeare describing an action of some kind in a scene from one of the plays. Sometimes a character like Iago acts as an enabler in the sense that the chopping and changing of his intentional activity mimes and reinforces Shakespeare's own. Shakespeare contrives a series of changes in the conditions of satisfaction of Iago's intentional action that follow upon changes in the way Iago is made to view the Intentional object at which he aims. Sometimes a plank is just a plank, in the same way as a 'monstrous birth' is just a 'monstrous birth'. But then a plank becomes Mr Smith's plank, and the 'monstrous birth' becomes putting the Moor 'At least into a jealousy so strong/That judgement cannot cure'. Iago stops short for the moment at this still fairly vague definition of his intention. In this respect he is unlike Macbeth, who moves at an early stage into an intentional position for which the condition of satisfaction is precise and (in a psychological, spiritual sense) terminal desideratum: Duncan's murder. From that point onwards Macbeth ceases to be an enabler of Shakespeare's enlargement of the field of intentional action in *Macbeth*. Iago, on the other hand, by having his actions alternate between more or less precise Intentional objects throughout the larger part of the play, is used by Shakespeare over a longer stretch of time to contribute to that sense of strenuous purposiveness without a clearly defined object that is so marked a feature of this play.

The way intentional actions are coaxed out of unintentional ones will have to be demonstrated on a micro-verbal level before it can be applied to the larger movements and changes in direction of the plays. But even here the agent of such a transformation must be the character working with the dramatist to enlarge the scope of his vocabulary and syntax so as to shift the centre of attention from one thing – a state of mind, a proposed course of action, a decision, a moment of reflection – to another – another state of mind, another proposed course of action, *etc* Shakespeare is always doing this sort of thing. Indeed it could be said that one of the main reasons his plays develop so urgently and forcefully and at so many levels of existential activity is that he is so adept at converting unintentional actions into intentional ones, and then discovering somewhere in the character of the newly created intentional action an unintended potential that will keep the play moving in ever-changing and unpredictable directions. This is one of the principal ways in which all those how-intentions reach towards but 'fail' to be swallowed up in the ultimate why-intention that is so persistently and energetically deferred. Scene follows scene, and always – or so it seems – the conditions for the satisfaction of the intentional action are never fully met, because the Intentional object is always changing.[34]

Take, for example, the 'To be or not to be' soliloquy near the beginning of the third act of *Hamlet*. Johnson says of it that it is 'connected rather in the speaker's mind than on his tongue', because it bursts out from the man who is 'overwhelmed with the magnitude of his own purposes'.[35] I want to show how these mental connections *are* reproduced in his speech, though in a manner unlikely to be appreciated by a critic like Johnson (for whom a quibble was a fatal Cleopatra), even when he notices the importance of multiple purposes in Hamlet's efforts to articulate his meaning. We need to see how one effort of this kind produces another, not in the shape of a logical précis such as Johnson supplies, but in the shape of shifting intentional actions of the sort I have been discussing in broader terms.

Johnson can make sense of the speech only by translating 'To be, or not to be' into a very definite statement of intent. He summarizes it as follows: 'Before I can form any rational scheme of action under this pressure of distress, it is necessary to decide, whether, after our present state, we are to be or nor to be.' In other words, he picks up the threads of Hamlet's argument about the existence or otherwise of a life after death from some time later in the speech, and uses it as an explanation of how the speech opens. This is good neoclassical practice. Johnson wants to bring out the fact of logical connection in the mind through a series of answering logical connections in the mode of expression. He does this by channelling the momentum of the speech in a single direction. Others have chosen to translate Hamlet's opening phrase in a different way, but with a similar purpose. Some editors have interpreted 'be' as an auxiliary verb preceding the 'understood' past participle 'done', with the result that Hamlet is seen to be meditating on his practical duty to kill Claudius.[36] Or they have played Johnson's trick of interpreting the words that follow 'question' to mean: 'I must take up arms against Claudius but that will probably bring about my own death too', and then shuffling them forward to a position in front of 'To be, or not to be', with the result that thoughts of dying follow closely upon thoughts of undertaking perilous future actions. Nevertheless, most critics assume they must start with thoughts of suicide and then make sense of the way such thoughts anticipate the speculations about acting or not acting that follow.

Bearing in mind the taxonomy of intentional actions sketched out above, we might be able to make sense of the connections in this speech in a more dramatically convincing way. The key to interpretation seems to me to be the use Shakespeare makes of a peculiarity of English grammar concerning the form of the verbal infinitive. It used to be said that the form of the infinitive in English is the same as the form of all verbs in the present tense with the exception of the third person singular, prefaced with the word 'to'. So: 'I eat', 'I dream', 'I act' are all functions of the verbs 'to eat', 'to dream', 'to act'. Really, though, this is misleading. It is true that the infinitive has an identical form with that of all verbs in the present tense with the exception of the third person. But the complementizer 'to' is added merely to distinguish it from other identical forms with different functions. It might be argued that it performs a residual grammatical function as a kind of suffix to the preceding finite verb, hence 'I want to → go', in preference to 'I want → to go'. Nevertheless we have fixed in our minds an idea of the infinitive that demands the 'to'-complementizer as an introductory particle, semi-detached from the rest of the word that articulates the mood of the verb.

This wouldn't matter in the ordinary course of things were it not for one or two other peculiarities of the English language. One of these is that the word 'to' also has the function of a preposition – two of the primary uses of which involve notions of having an aim or having an intention. *Webster* defines these uses as follows: '1. indicating the terminal point toward which movement is made or projected; as to drive *to* town; ten feet *to* the ground. . . . 3. indicating intention, purpose, or end; as hastened *to* our aid, title *to* property. . .'. The second is that 'to' before an infinitive sometimes does have a semantic function, which is related to *Webster* 3 above, i.e. in a sentence such as 'I took the knife to carve the joint', 'to' really means 'in order to'. Now this combination of grammatical peculiarities doesn't apply to any other language with which I am familiar. In French, for example, the different functions of the infinitive are clearly signalled by different prepositions. Hence in line 5 of Baudelaire's *L'Homme et la mer*, (*'Tu te plais à plonger au sein de ton image'*) '*à*' belongs to the verb 'plonger'; and in line 12 (*'Tant vous êtes jaloux de garder vos secrets'*) '*de*' governs '*garder*' in the sense of 'of guarding' (where in English we wouldn't use an infinitive at all). Usually the infinitive exists as a single word, with the inflexional '*er*' or '*ir*' etc to identify it. But where a sense of purpose is required, this is signalled by the prefatory preposition '*pour*': in Baudelaire again, see *Le Crépuscule du soir*, where the single-word infinitives '*siffler*', '*glapir*', '*ronfler*', '*commencer*', '*forcer*', terminate in the strongly purposeful '*Pour vivre quelques jours et vêtir leurs maîtresses*' – an effect it would be impossible to reproduce in English without inserting the awkwardly uncolloquial 'in order to' before the verb of purpose.[37]

How does this affect our reading of 'To be, or not to be'? The form of the infinitive here makes us attribute a strongly purposive function to an expression which needn't have any purposive function at all. The verb '(to) be' in English has a shadowy existence. We are used to hearing it either as an auxiliary forming the continuous present tense of another verb: 'I am going', 'he isn't taking me' etc, or as a sort of transparent pointing finger in indicative sentences of the type 'I am the Lord thy God', 'You are Peter', 'The Lord is my Shepherd', or as a shortened form of the expression 'has the property of' in descriptive sentences like 'You are clever', 'the grass is green'. In fact the verb tends to have very little verbal content, working instead to establish relations between other words in a sentence that are more important than it is itself. 'To be, or not to be' is a very odd phrase (only not sounding odd because we are so familiar with it in this passage). It has the effect of making the verb 'to be' purposive, and 'to be' is just about the most transparent, inactive and therefore unpurposive example of the infinitive in the English language. The tension between 'To' (which is a purposive, intentional word, implying activity) and 'be' (which is an unpurposive, merely formally functional word implying nothing to do with activity) is immediately felt. What Hamlet means by the words is something like 'A state of being, or a state of non-being', followed by 'that is the issue I would like to reflect on'. But in articulating it in this more verbal form, Shakespeare advertently (?) and Hamlet inadvertently make it sound less a subject for philosophical reflection and more an incitement to act in some way or other. And purposive action in relation to being, as distinct from all those other possibilities signified by more circumscribed 'doing' words (as they used to be called) can only be one of two things: continuing it, or putting an end to it. (Starting it is out of the question in Hamlet's present circumstances.)

Hamlet's 'intention to do' at the beginning of this speech is converted by the grammatical form Shakespeare chooses for him to speak it into an 'intention in doing' – which is that of making up his mind whether he should kill himself or not. In other words the conditions of satisfaction for the realization of Hamlet's intentional act are transformed in accordance with a shift in the character of the intention in action which is brought about by the particular form of words Hamlet uses. The unintentional act of deliberating on suicide immediately replaces the original intentional act of considering the merits of being and non-being in relation to the speaker's present state of mind. His own suicide then becomes the subject to which Hamlet devotes his attention in the rest of the speech. As Hampshire explains in a wider context:

> The representation of myself in words of an object desired modifies the direction, and sometimes the intensity, of the original, blind appetite. The desires develop and change direction in the process of elaborating in thought the possible ends of conduct, as in a Homeric council of war new courses of action are suggested and acclaimed.[38]

But because there is an emotional residue of the original intention still lurking in Hamlet's mind, the distinction between meditating and deciding tends to blur, especially when Hamlet uses infinitives (compare, for example, the form of 'to die to sleep' at line 64, with that of 'to die, to sleep' in the same position at line 60). By participating in the activities of Hamlet's mind, and reproducing their confusion in the language he is given to speak, Shakespeare opens up a range of possibilities for further development, just as he did by mystifying everyone, including himself in all likelihood, by making the status of the ghost so questionable in the first act. In both cases the discovery of intentional actions in descriptions of intentional actions – which has the effect of converting them, post-discovery, into alternative intentional actions – has a lot to do with the way the play moves purposively forward without disclosing a governing purpose. And the way the discovery is made might have as much to do with exploring the possibilities of grammatical form, or an image ('o'erleaping', 'engendering'), or a jump from one sentence to another, as with the conception and the positioning of a character, an event, or a twist in the plot.

This is how, in art, purposiveness can precede the existence of a purpose. By teasing how-intentions out of the most elementary generative potential – a ghost, a soldier, a prince, a word, an image, a sentence – Shakespeare gradually discloses the identity of a why-intention which was unknown at first to himself, and then, as they watched the play unfold, to his audience. To do that is to disclose a purpose which, though contained *in potentia* within the untargeted purposiveness of Shakespeare's imagination, was also, during the course of the play's unfolding, effectively unknown.

3
Coleridge and Kant

> ... so shall you hear
> Of carnal, bloody and unnatural acts;
> Of accidental judgments, casual slaughters;
> Of deaths put on by cunning and forc'd cause;
> And, in this upshot, purposes mistook
> Fall'n on th' inventors' heads...
>
> *Hamlet.*

SHAKESPEARE AND THE NINETEENTH-CENTURY NOVEL: PSYCHOLOGICAL AND INTENTIONAL MOTIVES

> The fated sky
> Gives us free scope, only doth backward pull
> Our slow designs when we ourselves are dull
>
> *All's Well that Ends Well.*

When the ghost appears to Hamlet in the queen's bedchamber (III. iv. 103 ff.), it tells him that 'this visitation/Is but to whet thy almost blunted purpose'. That purpose is to avenge his father by killing Claudius. It is a purpose imposed on Hamlet from outside, and we have seen the difficulty he experiences making it into something of his own. Shakespeare experiences the same difficulty over exactly the same thing – but on a dramatic rather than an existential level. For his purpose – imposed on him by the story of Hamlet he found in Saxo, Belleforest and Kyd – was also to kill the king. Like the ghost, he imposes this duty on Hamlet. When Hamlet has fulfilled his purpose in killing the king, Shakespeare will have fulfilled his purpose in writing the play about killing the king. There is a sense, then, in which the action of the play is regulated by a purpose not of the author's own contriving, but inherent in the narrative he has chosen to dramatize. Since this purpose can be completed only with the connivance of the principal character, the character must be instrumental to the author's purpose of satisfying the demands of the narrative he controls. We have seen, though, how Hamlet spends so much time evading his duty to satisfy his author's requirement of him. Looking at any episode in the play, we are likely to be struck by the way Hamlet's intentional acts fail to fit easily into a sequence of intentional acts which unambiguously look forward to killing Claudius as their *raison d'être*. The ulterior intention which will swallow them up in an explanation of what all that the play was for is as elusive as the ghost is enigmatic.

This aspect of the play fascinated Goethe, who discussed it in Book V of *Wilhelm Meister*. Before Serlo's troupe of travelling players begin to rehearse *Hamlet*, Wilhelm draws their attention to a discrepancy between the 'internal' and 'external' relations of the characters. Internal relations are 'the powerful effects which arise from the characters and proceedings of the main figures'. Wilhelm is perfectly satisfied with these. But the external relations – the means by which the characters 'are brought from place to place, or combined in various ways by certain accidental incidents'[1] – are unsatisfactory, and Wilhelm improves the play by subordinating them to a political subplot dominated by Fortinbras and the Norwegian business. This is what leads Wilhelm to his description of the play as one that is 'full of plan' but that has at its heart a hero who is 'without a plan'. The play he is referring to, however, is his own adaptation of *Hamlet*, not the play we know from the Quarto and Folio printings.[2] Wilhelm says that 'if the former background were left standing, so manifold, so fluctuating and confused, it would hurt the impression of the figures.' The puzzling concatenation of Chance and Fate in Shakespeare's version of the story invites a response pitched somewhere between the purely tragic and the proto-novelistic:

> The hero in this case, it was observed, is endowed more properly with sentiments than with a character; it is events alone that push him on; and accordingly the piece has in some measure the expansion of a novel. But as it is Fate that draws the plan; as the story issues from a deed of terror, and the hero is continually driven forward to a deed of terror, the work is tragic in the highest sense, and admits of no other than a tragic end.

This comes close to saying that the distinctive feel of *Hamlet* is attributable to the peculiar relation that holds between the hero's lack of a plan to determine what his actions will be, and the 'fluctuating and confused' character of the original plot. Wilhelm has to provide the play with a more coherent plot in order to emphasize the 'tragic end' to which the hero is drawn by circumstances over which he has little control.

In Shakespeare's play, Hamlet's lack of a plan is more subtly developed. Hamlet knows what the ghost's purpose is, but he has difficulty identifying it with his own purpose, though he knows that the two must in the end coincide. After all, it is only by their doing so that the play can end. Shakespeare also knows what the ghost's purpose is, and realizes it is identical with his own. But he shares Hamlet's difficulty, because it is his business not only to provide the play with an ending but to provide his ending with a play. Shakespeare needs Hamlet as both a help and a hindrance. He must fulfil the ghost's purpose and thus fulfil Shakespeare's purpose in using the 'Hamlet' story. He must *not* fulfil the ghost's purpose (throughout most of the play), and thus fulfil Shakespeare's purpose in holding the audience's attention for three hours' traffic of the stage. But there is more to it than that. The play could have been more like Kyd's – full of external impediments to Hamlet's execution of revenge. Shakespeare contrived that it should not be. He internalized the impediments and then dramatized them in the sequences of intentional action examined in Chapter 2 – where we found that the most conspicuous aspect of these sequences was their tenuous connection with what was inevitably to become the end of the play, i.e. what was demanded in some but not all of his moods by the ghost.

When Iago tells us about his purpose (in his soliloquy at the end of *Othello*, I. iii) he is referring to something that is in one sense similar to what the ghost mentioned in *Hamlet* and in another sense very different. It is similar in so far as Iago has to help Shakespeare perform the functions stipulated in Cinthio's plot. The story of *Othello* must end in Othello's death and, as in Cinthio, this must be brought about as a result of Iago's destructive purpose. The demands of the narrative force Shakespeare into using Iago as a way of ending the play by bringing about Othello's death. The difference lies in the fact that Iago invents his own plot to effect his own purpose. He needs no ghost to impose a purpose on him from outside.

When we listened to the ghost in *Hamlet*, I. v, we noticed how ambiguous his purpose was. The warning to remember and to revenge created the opportunity for Hamlet to procrastinate, substituting for the single intention – to kill the king – the multitude of intentions that prepared him for or deflected him from carrying out the ghost's purpose. In *Othello*, we noticed that Iago's 'purpose' in the soliloquy was merely formal. Iago uses the word to disguise the fact that he doesn't know what he intends to do. Nevertheless he is full of intentional energy. He has the will to perform malevolent intentional acts but, except in a very general sense (he wants to harm Cassio, Desdemona, Othello), he has no firm intention to turn that will into reality. The narrative thrust of the play works the way it does as a result of Iago's 'filling out' this intentional generative potential with more and more specific intentional acts that are finally swallowed up in his destruction of Othello and Desdemona. So his purpose eventually turns out to be their destruction. At I. iii, and for a long time after, he is not aware of this. His intentional acts gradually lead him to destroy the hero. But to assume that was his purpose at I. iii would be to credit him with clearer motives and longer-term intentions than he in fact possesses at this stage in the plot.

It doesn't matter that Iago uses a word like 'purpose' where what I think he means is a sort of cross between 'motive' and 'intention', any more than it matters in his second soliloquy (at II. i) that he uses a conjunctive phrase like 'For that' ('I do suspect the lustful Moor') or words like 'for' by itself (in 'for I fear Cassio with my night-cap too') when I think he means not so much 'because' as 'I intend to put it to myself that...' The main thing is to be clear in our own minds about what these words mean when we disengage them from the characters and apply them to the writer. We can appreciate Shakespeare's intentions, purposes and motives best – and understand their relevance to the responses we make to his work more clearly – if we have already discriminated among the different, but often closely related, concepts these words refer to. Now I want to release words and concepts from their specific applications to the characters' behaviour, taking each word separately and investigating the range of concepts it represents when we are talking about writing. Since I have earlier suggested a progressive scale of significance ranging from motive through intention to purpose, I shall begin at the least impressive, and in a way least relevant, end of the scale – with motive.

The concept of motive is an especially tricky one, because it can't help pointing in two opposite directions which are often confused with each other in textual interpretation. One way of thinking about motive is to ask 'why...' did it happen, did he do that, did he think one thing rather than another? This use of 'why?' refers to a prior occasion or state of mind that caused whatever came into being to do so. Hamlet's motive for putting on the *Mousetrap* might have been 'uncertainty' or 'a tendency to procrastinate' or 'because he mistrusted the ghost' or 'because he hated Claudius' or even 'because he loved the theatre'. Reading Victorian novels, interpretation of motive bulks large. Dorothea married Casaubon in the first book of *Middlemarch* for a variety of motives of this kind: because she was immature, because she was ardent, because she needed a father figure, because she wasn't clear in her own mind what marriage would entail, etc etc Notice that in asking questions about this kind of motive where fictional characters are concerned, answers the characters themselves couldn't be expected to supply are perfectly acceptable. Hamlet knew he loved the theatre but did he know he had a tendency to procrastinate? In a sense he did. Why otherwise would he have said what he said, or thought what he thought, in his soliloquies at II. ii, III. i and IV. iv? But did he know that was why he staged the play? The answer to this more specific question is not as easy to supply, and audiences will differ in their assessments of the extent of Hamlet's knowledge or ignorance of such a motive as applied to such an activity. Dorothea knew she admired middle-aged scholars, but did she know she was ardent? In a sense she did. But not in the same way as she knew she admired middle-aged scholars. And did she know she was confused about what marriage would entail, or that she needed (thought she needed?) a father figure? This would require a much more complicated answer, hedged about with many more qualifications. Hamlet and Dorothea know little about this kind of motive for their behaviour – less, in fact, than we are led to believe we do ourselves.

There are many occasions in European literature when readers have felt inclined to ask questions about motive in middle- or higher-order ranges of interpretation, even though the status of the events about which the questions are asked is ambiguous. In Henry James, for example, they are legion. Why does Isabel Archer return to Gilbert Osmond at the end of *The Portrait of a Lady*? Why is it impossible for Merton Densher and Kate Croy to marry at the end of *The Wings of the Dove*? Why does the Princess send her father and Charlotte away to American City in *The Golden Bowl*? The question about *The Portrait* suggests that 'to return to Gilbert Osmond' is an alternative to other courses of action Isabel is at liberty to choose from. It conceals the pre-supposition that there is a decision to be made one way or another. True, the novel does attribute to Isabel a number of possible motives for returning or not returning. These are of the 'reply-to-why?' variety we are considering now, and also of the other sort we have not yet discussed. But if we were to put to one side this question about a middle-order issue, and for the time being substitute for it some question about a prior, and lower-order issue of the 'why-does-Hamlet-speak-these-particular-words?' variety, we might discover that we can no longer put that earlier question to ourselves in quite the same way. We could turn to the beginning of chapter 55 of *The Portrait* and ask why Isabel speculates about 'certain obligations [that] were involved in the very fact of marriage'. James describes Isabel's thoughts in the free indirect style, making her the subject of his third-person narrative, but using the words she would use to imagine the position she thinks she occupies: 'She lived from day to day, postponing, closing her eyes, trying not to think. She knew she must decide, but she decided nothing; her coming itself had not been a decision. On that occasion she had simply started.' And so on. Deciding not to decide indicates a quite different frame of mind from one about which one might ask the question: why did she do... whatever it was she did? – in this case, return to her husband. The terms in which she puts to herself what most matters to her prevents the reader from formulating a description of her situation as one in which she makes a decision to do something. The question is not strictly relevant to Isabel's state of mind at the time she is understood to be making that decision.

The reader is tempted to avert his eyes from the precise words James uses to represent Isabel's turn of thought, substituting for them a higher-level description couched in familiar motive-terms that transform it into a manageable 'event'. All James's prose has allowed Isabel to think to herself is simplified into the stereotypical motivational terms of decision-making that make 'why did she do... what she did?' an appropriate and straightforward question to ask. Now it becomes a question that can be answered in terms of a prior emotional predisposition of a readily identifiable kind, or, more misleadingly still, in terms of an aim to be achieved (the answer to a why-question that takes the form of 'because she wanted to achieve these particular results').

We are likely to find ourselves answering this second question because descriptions of states of mind have a habit of sliding into descriptions of mental functions which propose to themselves an aim. Descriptions of Isabel's emotional predisposition at the point where she hovers between the rival claims of England and Italy are bound to keep spilling over into the matter of what she proposes to do. Consequently questions about motive (why, i.e. in accordance with what emotional predisposition, does she behave in this way?) imperceptibly transform themselves into questions about aims (for what reason, with what purpose or end in view, even with what intention, does she behave like that?). It is easier to describe Isabel's motive in terms of what she aims to achieve by doing what she does than in terms of what it is about her emotional life that triggers off such an activity. It is easier to say: 'She wants to protect Pansy' or 'She feels the need to keep up appearances', or even 'She wants to provide Osmond with all that he could have legitimately expected to have gained by marrying her', than it is to trawl through Isabel's past and bring to the surface the multitude of relations, responses, decisions and speculations that have prompted her to act the way she does.

The same is true of the characters in *Hamlet* and *Middlemarch*. To the question 'why does Hamlet disbelieve the ghost?' (or half-disbelieve the ghost, or pretend to half-disbelieve the ghost) it is easier to answer 'because he doesn't want to take action against Claudius' or 'because he knows that if he doesn't he will have to do something practical about what he takes to be Denmark's rottenness and woman's frailty', than it is to answer by referring to the complex state of mind which accounts for his taking this sort of view of hypothetical future actions. Or to the question 'why does Lydgate call on the Vincys in chapter 31 of *Middlemarch*?' one is more likely to answer 'Because he wants to see Rosamond' (followed by suitable qualifications) than to go into all the details about Lydgate's immediate and not so immediate past that accounts for other motives he might have had for calling on her.

In all these cases there is a motive involved in the character's contribution to middle-order events that moves us forward into hypotheses about later, more comprehensively explanatory events of the higher-order kind. And there is another, related motive that puts us in mind of earlier, more narrowly explanatory events. The first motive approximates closely to what we often mean by intention: Isabel intends to preserve her marriage; Hamlet intends to delay taking action against Claudius; Lydgate intends to satisfy his curiosity about Rosamond. To the extent that it does this it seeks to merge with the highest-level aspect of the play, novel, story that has anything to do with motive. Isabel gradually comes to understand that her fate is to preserve her marriage to Osmond. Hamlet understands more and more clearly that his fate is to kill Claudius only after fabricating opportunities for maximum delay. Lydgate sees that his fate is to have his most cherished ambitions thwarted by his marriage to Rosamond. None of these characters knows at this point that the effect of his having these motives, i.e. 'having aims for which there are reasons', will be to draw him towards what it is his fate ultimately to be or do. Because the characters don't know the ultimate intention their present behaviour will eventually lead them to possess, their motives for what they do can be described as intentional only in the sense that by performing A they can foresee that they will inevitably bring about B or C. The why-intention that will finally swallow up the earlier how-members of the series is for the moment inconceivable. So far as the character is concerned, B or C is the final why-intention. He cannot foresee the end of the story. And neither can the writer – in any but the most superficial and mechanical sense of foreseeing.

The other class of motive – or rather the same motive looked at from the other side of the 'why?'/'what for?' interpretative schema – doesn't approximate to any meaning of 'intention' at all. Since it doesn't look forward to a hypothetical future, it doesn't set in motion any intentional acts that might have the effect of turning the hypothesis into a reality. Because it has no immediate link with future activity, it is much more difficult for the author to describe or for the character to know. In fact the descriptions are, tacitly, what we have been told about the character, in often seemingly gratuitous detail, at earlier points in the narrative. The character's knowledge of what makes him act as he does is usually incomplete or even mistaken – in a way that he cannot be conceived to be unknowing or mistaken about his intention (in the sense of what it is he proposes to do).

In what follows I shall differentiate between two kinds of motive by naming the first 'psychological' and the second 'intentional'.[3]

Psychological motives are always retrospective. Although the 'matter' out of which they are formed is experienced in the reading as present and actual fact, by the time we come to consider this 'matter' as motivationally significant it has become part of the past and is seen to be so by the reader and by the character – if he is the sort of character who speculates about such things. Obviously this must be the case as far as background is concerned. But it is true of disposition also. For at the point at which disposition, attitude or temperament has entered our or the character's field of speculation about motive, it has become something-in-the-present-which-has-come-out-of-the-past. Once it changes from that into something that exists in the present but makes a character look forward to the future, it has stopped being a psychological motive and become an intentional one. At this point the psychological motive becomes irrelevant from the character's point of view. Absorbed in speculation about the future, he breaks the psychological connection with past events and attitudes that have led him to this activity. So intentional motives are always anticipatory. They are also voluntary, because, unlike psychological motives, they seek to impose themselves by converting hypothetical events into real ones. But the moment they begin to do this, they stop being motives and become acts. In order to have the attribute of being acts that follow on the character's consciously (and there is no other way) being in possession of an intentional motive, these acts themselves must be intentional.

Anscombe (pp. 18–19) sees little difference between motive and intention.[4] She says that although in philosophy a distinction has sometimes been drawn, as if our motives and our intentions in acting were quite different things, really this comes close to being a distinction without a difference. 'A man's intention is what he aims at or chooses; his motive is what determines the aim or choice.' In Chapter 2 above, taking my cue from Anscombe, I referred to the difference as one between the elliptical and non-elliptical forms of, for example, 'gain' in a context of assessing a businessman's or a gambler's intentions and motives. Where 'gain' is the intention, 'desire of gain' is the motive. Goldman makes a similar distinction between cases in which the goal of an agent in performing an action seems to be the possession of an object and those in which the goal is the performance of an act: 'I give the automobile dealer a check because I want to own a certain car; but owning the car is not an act.'

In this case 'I give the dealer a check because I believe that this will generate the act of my gaining ownership of the car, an act I want to perform.'[5] But as Anscombe implies, most people would accept 'gain' as a motive for transacting business, using it interchangeably with 'desire of gain'. 'Gain' is something that someone who is motivated by 'desire of gain' hopes or predicts will come into being as a result of his acting in accordance with that motive. It doesn't exist apart from the activity that takes it as its object, or from the conclusion of that activity in having accomplished what it set out to do. So while Anscombe's 'desire of gain' remains a motive in both her and my vocabulary (a psychological motive in mine), 'gain' is an intention in her vocabulary, but both an intentional motive and an intentional act in mine. The word functions in the same way as many abstract nouns do that have a pronounced verbal aspect. It isn't used interchangeably with 'desire of gain' alone. It is used with the 'idea of gaining' and 'the act of concluding gaining something' as well. As 'the idea of gaining' it is an intentional motive; as 'the act of concluding gaining something' it is an intentional act. Intention itself is a mysterious nonentity. No wonder theoretically sophisticated critics have so easily dispensed with it in their approach to literature and art.

Wimsatt and Beardsley seem to think of intention as a real state of mind the writer possesses before he puts pen to paper. Then it miraculously disappears from view. Who cares, they say, that Coleridge read Bartram's *Travels* some little time before writing 'Kubla Khan'? He might have done that and, having done it, formed the intention of using his reading to contribute to the creation of the poem, even if this intention was formed 'in the deep well of Coleridge's memory'. But 'it would seem to pertain little to the poem to know that Coleridge had read Bartram':

> There is a gross body of life, of sensory and mental experience, which lies behind and in some sense causes every poem, but can never be and need not be known in the verbal and hence intellectual composition which is the poem. For all the objects of our manifold experience, for every unity, there is an action of the mind which cuts off roots, melts away context – or indeed we should never have objects or ideas or anything to talk about.[6]

Wimsatt and Beardsley are not really talking about intention here at all. How could they be when, apart from intentional motive and intentional action, such a thing can scarcely be said to exist? What they have in mind is psychological motive, interpreted in the widest sense to subsume all the significant elements of the writer's biography – including his reading. This is something quite different from what they announced was their subject – 'the design or intention of the author' – in the first paragraph of their essay. Different also from the 'design or plan in the author's mind' which they defined intention as before they went on to adduce inappropriate examples of such intentions in Coleridge, Donne and Housman. Somehow they have shifted their attention from the most conscious and deliberate form of intentional activity applied to a specific aim (the aim of writing the poem) to a more amorphous combination of activities existing at a greater remove from the poem than anything involving 'design' can do. Replying to Wimsatt and Beardsley, we have to begin by distinguishing between the psychological motive they are in fact discussing for most of the time, and the combination of intentional motive and intentional activity that is the proclaimed but not the actual subject of their essay. Dispensing for the time being with intentional motives – which do seem to lead the critic into murky waters – let us turn to the two conditions or activities that 'intention' has separated into: intentional motive and intentional acts.

I have defined intentional motive as the motive a person attributes to his actions when asked the question 'what for?' In colloquial speech 'what for?' often performs the same function as 'why?', and vice versa. The two words/phrases have closely related semantic functions. When they are used to ask questions about human behaviour both of them manifest a curiosity about motive. But the one asks about psychological motive, the other about intentional motive. We are asking different things when we ask 'why did Hamlet kill Claudius?' and 'what did Hamlet kill Claudius for?' The answer to the second question might be 'to gain the throne of Denmark' or 'to avenge his father's murder', whereas the answer to the first would be something like 'because he loathed him', or 'because he couldn't see why he shouldn't perform a pointless act in the pointless world to which he considered his creator had consigned him'. Either answer could be simple or complex, but the answer to the second question will tell you something about the character's intentions, whereas the answer to the first will tell you something about his predisposition or attitude of mind.

An intentional motive is the motive for doing something conceived of in terms of what it is done *for*. If I am stooping outside my front door with a spade in my hand, shovelling snow away from the path, I might well be asked by a lazy neighbour what I am doing that for. My answer will be as follows: I'm doing it so that I can get my car out of the garage; or, more informatively, I'm doing it so that I can get to the bank before it closes; or, more informatively still, I'm doing it so that I can get money from the bank to pay the builders. In this case the intentional motive fully explains the intentional act that expresses it. There wouldn't be much point in drawing a distinction between them. Also, there is another respect in which two ways of speaking about intention would come to just about the same thing here. By shovelling the snow, I *intended to* clear the path for my car, enable myself to get to the bank, make it possible for me to pay the builder – in other words perform all those intentional acts which are ultimately swallowed up in the act of paying the builder. Also, my *intention in* shovelling the snow was exactly the same: to get to the bank, get out the money, pay the builder. But is this always the case? More especially, is the identity of *intending to* do something and *having an intention in* doing something preserved or not preserved when one moves from considering a simple activity like paying the builder to considering a more complex activity like killing the king, returning to Gilbert Osmond or marrying Rosamond Vincy?

It will simplify the business of writing about the issue if we substitute Anscombe's 'intention of doing' a thing for my 'intent to do' it, preserving its relation of identity with 'intention in doing' what is done wherever simple acts are concerned. Intentions *of* doing and intentions *in* doing might boil down to exactly the same thing from a practical point of view – when the object of the inquiry is a simple matter like shovelling snow. But the practical similarity conceals a theoretical difference which might have a noticeable effect when questions of the same sort are asked about more complex activities. For example, Iago's intentions of doing what he set out to do in *Othello* are very different from his intentions in doing what he does in fact do.

My state of mind when I am about to go out and shovel snow from the path is uncomplicated. I realize I need to pay the builder. To pay the builder I need some money. To have the money ready I need to get it from the bank. To get to the bank I need to use the car. So here I am in the present, looking towards an event in the future (paying the builder), and proposing to myself a course of action that

Coleridge and Kant

leads back from putting the money into the builder's hand to grabbing hold of a spade to shovel the snow. D is related to A in as simple a reversed sequence of effects and causes as poisoning the fascists was related to the man's moving his arm up and down in Anscombe's fable. The man intended to poison (had the intention of poisoning) the fascists. His intention in doing what he did (moving his arm up and down, *etc*) was that of poisoning the fascists. It comes to the same thing. Similarly, I intended to pay the builder. My intention in shovelling the snow, *etc* was that of paying the builder. The only difference between the two activities is that in my activity there is a wider temporal spacing between causes and effects. Shovelling the snow prepares for driving out the car which prepares for... *etc* But moving my arm up and down doesn't just cause the water to flow; it is another description of doing that – in the context of this particular sequence of activities. However, this is a trifling difference within the context of my present argument. The much more important similarity lies in the clear and uncomplicated relation between an intention it is proposed to realize and a series of closely connected actions I envisage will result from its having been realized.

What happens when we move away from the path and the spade and return to Venice and Iago? What is the equivalent of my wanting to pay the builder, i.e. what I had the intention to do? We should be able to answer this question by referring to Iago's soliloquy at I.iii. where he says he intends to get Cassio's place and plume up his will. How does he propose to achieve what he intends to do? By abusing Othello's ear. That is the equivalent of my shovelling snow and driving to the bank. What am I shovelling snow for? Ultimately, to pay the builder. What is Iago abusing Othello's ear for? Ultimately, to get Cassio's place and plume up his will. He intends to get Cassio's place, and his intention in abusing Othello's ear is to get Cassio's place. On this occasion I shall ignore the complication of the second, less palpable intention (pluming up his will). Accepting, though, that Iago's intention is as definite and comprehensible as 'to get Cassio's place', isn't there still a real difference between his attempt to realize this intention, (in putting to himself how it might be achieved – by abusing Othello's ear, *etc*), and my realizing my intention (in putting to myself how it might be achieved – by shovelling the snow, *etc*)? Surely the difference has to do with the degree of congruity or incongruity in the relation of means to ends? What we are speculating about is the presence or absence of links in the chain connecting the formulation of an intention (having an intentional motive), the performance of a series of intentional acts which are expected to realize what was intended, and the actual realization of what was intended through the agency of these intentional acts.

When I formulated my intention of paying the builder, I traced a course of action backwards from the point at which I put the money in his hand to the point at which I picked up the spade. Any time that elapsed between my formulating this course of action and picking up the spade was insignificant from the intentional point of view. To pay the builder I simply performed the intentional acts I had earlier put to myself as hypothetical acts that would become real as and when they were performed – though in temporally forward rather than reverse order. This is what Iago doesn't do in the soliloquy. He must be presumed to have done it elsewhere, because he does have a plan to disgrace Cassio by getting him drunk when he is on the watch, and we are to suppose that the reverse-order hypothetical intentional acts in Iago's planning were: I get Cassio's place ← Othello dismisses Cassio ← I get Othello to witness the drunken brawl ← I provoke the drunken brawl ← I persuade Cassio to have too much to drink. Abusing Othello's ear therefore takes the form of lying to him about Cassio's drinking. Here we have a sequence of intentional acts not at all dissimilar, except in their dramatic interest, to my shovelling the snow in order to pay the builder.

The difference lies in the fact that the sequence of intentional acts which fell between my intentional motive of paying the builder and my actually paying him was complete. There was an unvoiced assumption that the builder had finished his building and I had no further plans for anything to do with the building. The notion 'I must pay the builder' and the events that were consequent upon my having that notion, terminating in my paying the builder, comprised the whole of the 'play'. This is by no means the case with Iago. For a start, Cassio is persuaded not to accept his dismissal tamely, and in any case a lot happens between his being dismissed and Iago's actually getting his place. This is quite different from the snow scene, where the builder was presumed to be satisfied with his payment and it wasn't a bank holiday. Well, it might be argued, even if these complications had arisen, that wouldn't significantly alter the connections I have been describing between having intentions, doing things in order to realize them, and then realizing them. This would be true if the attitudes and dispositions of the players in these little scenes remained stable. So far I have been tediously public-spirited, the bank has been doing what banks are supposed to do, the builder has patiently waited to receive his money. But what if I get angry at the bank's being closed, the bank manager is drunk and disorderly after his bank holiday celebrations, the builder is disposed to haggle about his price? Or, in *Othello*, so far Iago has been single-mindedly devoted to the task of pulling Cassio down a peg or two, Othello has been duly grateful for Iago's show of loyalty, and Desdemona unconcerned about Othello's relationship with Cassio. But what if Iago discovers that he hates Othello more than he envies Cassio, Othello becomes irritated at Desdemona's representations on behalf of his sacked lieutenant, Desdemona naively presses for Cassio's reinstatement? In each case the likelihood is that the intention will change along with the circumstances of the people involved. The intentional motive becomes not so much to pay the builder as to argue with him about his price, and this initiates a series of intentional actions which are only tangentially related to the original intentional motive of paying the builder. We have already seen that Iago's success in getting Cassio demoted drives him beyond the satisfaction of his original intentional motive, to engage in very different and more far-reaching intentional acts which ultimately bring about the wholesale destruction of the dramatis personae.

The issue becomes more complicated when we reinsert Iago's phrase about pluming up his will. The presence in Iago's mind of two intentional motives, which may or may not easily coexist, and which may or may not, therefore, issue in the same or similar intentional actions, does significantly alter our appreciation of what Iago intends to do and what his intention is in doing it. For each stage in the subsequent action of the play, we will have to ask ourselves not just what is Iago intending to do, but in respect of which of these two intentional motives is he intending to do it? Looking forward from the soliloquy we would like to envisage one of two things: either a single course of action that is consequent upon the two intentional motives being more or less the same, or at any rate the one motive entailing the other; or two courses of action moving forward concurrently but, because they derive from separate intentional motives, never at any point impinging on each other in a way that creates awkward interpretative complications.

I say we would like to envisage one of these two things. And this is true, I think, if we are talking in commonsensical terms. If, though, we are speaking in 'theatrical' terms, what we are likely to envisage will probably be the very opposite. The satisfaction we gain from watching *Othello* has more to do with our pleasure at the confusion of intentional motive in relation to intentional acts than with the pleasure we would take in other circumstances in the uninterrupted sequence of intentional motive → intentional act → satisfaction of intentional motive in the way all the other members of the series of intentional acts are swallowed up in its final member. Iago's intention of doing the things he proposes to himself turns out to be something very different from what we see are his intentions in doing some of these things (and others) at the moments when he does them. Looking forward from the intention of doing, we think we see a final intentional act to which everything that follows will lead. But when the intention of doing is set in motion, it loses its monolithic and narrowly motivated character. Iago's intention in doing what he does exposes complications that were latent in the phrases he used in the soliloquy. As a result of this, the momentum of the play is interrupted, and the direction we expected it to move in is deflected – on increasingly frequent occasions as events multiply and the characters' involvement in them becomes more intricate. As this happens, the expectation we might have had of a perfect fit between motive and result, aim and target, is disappointed. But by the end of the play something has happened to give it a shape, a point, a purpose after all. Looking back from the last scenes, we can see that some sort of end has been accomplished, some kind of purpose fulfilled – by Iago and the other characters on Shakespeare's behalf. What, then, is the link between what Shakespeare intended and what his characters intend? In what senses can what they do be described as Shakespeare's intentional acts? Do a writer's intentional acts also outstrip the intentional motives with which he sets out to perform them? And how does all this relate to my rather abrupt use here of the third of my terms, purpose, in describing the momentum and direction of the play?

COLERIDGE ON KANT

> Why should we suppose that nature acts for something...? Why shouldn't everything be like the rain?
>
> <div align="right">Aristotle, <i>Physics</i>.</div>

We talk of someone doing something for a reason or for a purpose interchangeably, meaning either that he has a definite aim in mind or that he has a comprehensible motive for doing what he does. The first meaning is closest to what the word 'purpose' usually suggests and to what it is required to suggest in the present argument. 'Purpose' in this sense has so much in common with an intention to do something, or the intention of doing something, that the words and phrases are practically identical. The assumption tends to be that whatever that 'something' is, it has a definite content, which is an object in the mind of the person who has the intention or purpose reaching out to it. The sense we have that to be purposive is to be directed towards some specific and definite accomplishment is evident in the phrase '[to do something] of set purpose', where 'set' includes the meanings of both 'settled' and 'prescribed'.

'Purpose' is another way of describing an intention of doing. As such it is to be distinguished from 'purposiveness', which sounds much closer to it than in fact it is. One of the problems here is that the 'purposive' is used as the qualifying word derived from both of these abstract nouns. But it is crucial to the success of the argument here that we distinguish between purpose and purposiveness, as between a mental inclination that is on the one hand definite, specific and scrupulously motivated, and on the other hand one that is indefinite, tending to more general aims and unregulated by any specific motive or group of motives.

That is what Kant suggested in his *Critique of Judgement* when he used the words '*Zweck*', '*Zweckmässigkeit*' and '*Endzweck*' to differentiate between comparatively closed and open descriptions of purpose-laden activities. In what follows I want to review Kant's distinction between purpose and purposiveness, bringing it into conformity with the distinction already made between different kinds of intentional activity. For it will be apparent that what Kant means by purposiveness has as much relation to an aggregate of intentions of doing as what he means by purpose has a relation to particular intentions of doing. If this can be shown to be the case, an argument about a single (though controversial) aspect of the psychology of the imagination might be developed into a wider argument about the basis of aesthetic judgement. That is what I propose to enlist the services of Kant in order to achieve.

The history of Kant's reputation in Europe is to a large extent synonymous with the history of German Idealism from the early years of the nineteenth century, through the dialectic of Hegel, to the phenomenology of Jaspers and Heidegger in our own century.[7] In England, though, the situation is different and altogether more tractable. For historians of philosophy, the transformation of the Kantian critical philosophy into its Hegelian apogee in the later years of the nineteenth century is all-important.[8] But for students of literature, and of the theory of literature, it is clear that Samuel Taylor Coleridge is the central figure. Coleridge was the poet and thinker who dictated the way Kant's aesthetic philosophy would be understood in this country. His was the mind that Idealized the Kantian philosophy of mind and art, thus establishing the parameters within which Anglo-Saxon writers sought to define the workings of the imagination and the status and nature of the objects it constructs. To understand how we have accustomed ourselves to view notions of intention and purpose in literature it is therefore necessary to study Coleridge's literary theory, and especially that part of it – the core of it, as I shall argue – that owes most to the example of Kant.

In the ninth chapter of the *Biographia Literaria*, Coleridge makes his most celebrated reference to Kant's influence on his thinking: 'The writings of the illustrious sage of Konigsberg, the founder of the Critical Philosophy, more than any other work, at once invigorated and disciplined my understanding.'[9] Now, in 1817, Coleridge refers to 'fifteen years' familiarity' with these writings, which takes us back to the very early years of the century, shortly after the first publication of Wordsworth and Coleridge's *Lyrical Ballads* and the poets' return from their visit to Germany of September 1798 – July 1799. It is worth while asking how accurate Coleridge is in tracing his familiarity with Kant's three *Critiques* to these dates.

The first reference to Kant in Coleridge's writings is on 5 May 1796, in a letter to Thomas Poole making plain his intention of reading Kant.[10] Clearly he had not yet done so. In a letter to John Thelwall in 17 December of the same year, he implies that he has read Kant, but finds his philosophy 'unintelligible'.[11] Here as elsewhere in his writing near the turn of the century the references are extremely vague. It is not easy to disentangle them from his speculations about German metaphysics in general. The same is true of his four 'metaphysical letters' to Josiah Wedgwood of February 1801,[12] in which it is difficult to measure the impact of Kant on his reading of Descartes and Locke since Kant's name is never mentioned. In his letter to Poole the following March his comments on Space and Time appear to owe something to the *Critique of Pure Reason*,[13] but the debt is neither precise nor acknowledged. Rosemary Ashton[14] dates Coleridge's first reading of Kant to 1801, and there is general agreement that he undertook what the editor of his *Marginalia* describes as an 'intensive reading of Kant' in the winter of 1800–1. Even so, references to the *CPR* specifically 'do not occur before 1807–8'.[15] References to Kant's other works are equally scant. Ashton agrees that there is little hard evidence of specific Kantian influence on his writing at this time – in the poems, letters or even in the notebooks.

The earliest comment on any specific aspect of Kant's philosophy in the correspondence is in a letter to John Ryland, 3 November 1807: 'Kant in his Critique of Pure Reason, and more popularly in his Critique of Practical Reason, has completely overthrown the edifice of Fatalism, or causative Precedence as applied to action.'[16] Wellek argued that Kant's influence on Coleridge first became noticeable during the writing of *The Friend* (1809–10),[17] but I can find little evidence of this in the essays themselves. On the other hand there are Kantian distinctions between pure Reason and the prudential Understanding in letters written by Coleridge immediately before and during publication of *The Friend*. However, we have to wait until some time after 1807–10 to uncover evidence of his more intimate acquaintance with the *Critiques*. Coleridge appears to have reread Kant in 1809 (while editing *The Friend*), 1817–18 (*after* writing *Biographia Literaria*) and in the mid-1820s (exerting a profound influence on the argument of *Aids to Reflection*). These deductions scarcely support his claim in the *Biographia Literaria* that his reading of Kant during the late 1790s was so extensive and its influence so powerful as to have 'at once' transformed his understanding of philosophy and literature.

Another interesting feature of these references to Kant is that all of them are to the first two *Critiques*. None suggests familiarity with *CJ*. We have to remember that the third *Critique* was published in German as recently as 1790, that English readers had to manage with poor French translations until J.H. Bernard's translation into English as late as 1892, and that Thomas Beddoes' essay on Kant in the *Monthly Magazine* (May 1796), which we know Coleridge read with enthusiasm, included material only on the first two *Critiques*. Even so, Coleridge was proficient enough in German by the early years of the new century to read *CJ* in the original, and we know he possessed a copy of the 1799 edition of the German text. Ashton quotes a passage from a letter to Sotheby in 1802 as an 'early' example of Coleridge's use of Kantian distinctions between nature and spirit which she suggests he might have found in *CJ*.[18] The argument is unconvincing. Especially before his first 'rereading' of Kant in 1809, but to an unprejudiced mind on most occasions thereafter, Coleridge's references to the third *Critique*, in the very few contexts in which they are discernible, are even less specific than are those to the two earlier ones.

According to the editor of Coleridge's lectures at the Royal Institution in 1808, there are early echoes of *CJ* in the notes he wrote for this occasion.[19] The editor of the Shakespeare criticism had already drawn attention to the most impressive of these echoes – the one from Lecture 4 about 'catch[ing] a Hint' of art 'from nature itself '[20] – so the accumulation of evidence does point to Coleridge's having acquired some knowledge of *CJ* at some time before or during 1808. But it is a very imperfect and imprecise knowledge. Most of it derives from Kant's distinction between natural and artistic beauty, which was probably the part of his work most widely circulated in the early years of the new century. The notes to Lecture 14 have not survived, but Crabb Robinson attended it and wrote an account to Mrs Clarkson which tells us that 'Coleridge contrived to work into his speech Kant's admirably profound definition of the naif, that it is nature putting art to shame.'[21] Wellek looks for evidence later than 1808. He directs us to the fragment of the *Essay on Taste* of 1810, confessing that it reveals 'little more than the beginning of a statement of the central problem of the *Critique of Judgement*'.[22] Certainly it would require only the most desultory reading of *CJ* to define taste as the ability to 'combine and unite...a sense of immediate pleasure in ourselves with the perception of external arrangement'. This is so vague as to have been derivable from the work of almost any English or German writer on aesthetics since the beginning of the eighteenth century.

Much the same applies to Coleridge's comments on Kant in the Shakespeare lectures of 1812–13, in which, according to Crabb Robinson, 'He [Coleridge] enlarged on the vagueness of terms and their abuse and in defining taste gave the Kantian theory as to the nature of the judgments of taste.' His editor comments that 'Robinson is rather vague and his reference is sufficient to involve the entire first book ["The Analytic of the Beautiful"] in Kant's "Critique of Aesthetic Judgment",'[23] but there is no reason for us to suppose that this involvement was very detailed. Payne Collier's shorthand notes from these lectures are more detailed than Robinson's of the 1808 series, but they are neither complete nor entirely reliable so we simply cannot say. Incidentally, none of the plagiarisms and paraphrases tabulated by Snyder in her edition of the 'Logosophia' notes Coleridge composed during most of his mature writing life are taken from *CJ*.[24] So Coleridge's claim, in a letter to an unknown correspondent between 15 and 21 December 1811 to have 'mastered the spirit of Kant's Critique of Judgment'[25] seems fragile. It is unlikely that by 1808, or even 1812, Coleridge had more than the skimpiest knowledge of *CJ* or any but the faintest recollection of his earlier reading of it back in the late 1790s or early 1800s.

It was in 1814, with the essay *On the Principles of Genial Criticism*, that the Kantian reference became more precise. Here taste is defined as a mental faculty mediating between other mental faculties much as the judgement does in *CJ*. 'Taste', Coleridge writes, 'is the intermediate faculty which connects the active with the passive powers of our nature, the intellect with the senses; and its appointed function is to elevate the images of the latter, while it realises the ideas of the former.'[26] In a book containing much on the subject of Coleridge's borrowings from Kant, Norman Fruman suggests that 'Throughout the first two essays [of *Principles*] the discussion of beauty draws heavily on Kant's analysis of beauty in the *Critique of Judgement*. At all times, however, the concepts are advanced as completely novel.' He points to 'numerous and detailed borrowings from the *Critique of Judgement* in these essays...'[27] Coleridge writes:

> The Beautiful arises from the perceived harmony of an object, whether sight or sound, with the inborn or constitutive rules of the judgement and the imagination: and it is always intuitive. As light is to the eye, even such is beauty to the mind, which cannot but have complacency in whatever is perceived as pre-configured in its living faculties.

James Engell also emphasizes the *Principles*' debt to *CJ*: 'Coleridge insists ... that beauty in art is thus a calling on the soul, and – drawing on Kant's *Critique of Judgement* – that beauty is an ultimately intellectual satisfaction perceived through the mediate senses, but not belonging to them.'[28]

The terminology and general picture of the mind are Kantian. The purposes to which the terminology is put, though, and the particular kinds of mental activities described, are significantly unlike the ones Kant describes in *CJ*. However, they are sufficiently like them to cause confusion. Coleridge's use of faculty psychology, for example, is like Kant's, and the description of taste as a faculty mediating between two 'powers of our nature' sounds as if it derives from *CJ*. The identification of these powers as 'intellect' and 'sense', however, and the ascription to taste of the capacity to realize intellectual ideas, whether or not in association with images, has no warrant in Kant. Nor has what Coleridge calls 'pre-configuration' of the objects of perception and the living faculties of the mind that perceives them according to 'inborn and constitutive rules'. As we shall see, there is a great deal of uncertainty about the degree to which indeterminate concepts (not rules) proposing some sort of fit between mind and nature can reasonably be described as constitutive. For Coleridge, however, it is important that they should be so described. The argument anticipates that of chapter 9 of *Biographia Literaria*, although there the Kantian terminology (at any rate of the *CJ*) has been dropped at the same time as the acknowledgement of Kant's influence has become more explicit than it was in Coleridge's writing earlier in the century.

The most suggestive feature of Coleridge's references to Kant in *Biographia Literaria* is the discrepancy between the list of books by Kant he says he has read and the ideas he claims to have discovered in them. In this list, the *Critique of Judgement* is named between the *Critique of Pure Reason* and the *Metaphysical Elements of Natural Philosophy*. 'Fifteen years of familiarity' with this book seem to have left only the faintest impression on Coleridge's mind, a fact that appears all the more remarkable when one considers that *CJ* devotes more than half of its pages to the subject of judgements of taste and the nature of the beautiful and the sublime. Nowhere does Coleridge mention Kant's discussion of the relation between the Form of finality, or purposiveness, and judgements of taste – the very heart of his aesthetic. Instead, either he writes in very general terms about the originality and depth of Kant's thought and the 'adamantine chain' of his logic (which, amazingly, he thinks is 'clear' and 'evident'), or he considers in greater detail such concepts as the categorical imperative and the *noumenon*. These are not concepts with which Kant is much preoccupied in *CJ*, certainly not in the part of it ('The Analytic of the Beautiful') that is mainly concerned with aesthetic matters. In so far as Kant seeks to apply them at all to his aesthetic, he does it in his consideration of the sublime.

One might argue it is likely that Coleridge, a Romantic poet and critic, would be drawn to this aspect of Kant's work. The fact remains that what Kant says about the sublime has to do almost entirely with aesthetic approaches to nature, not art, and that in any case he doesn't argue that those supersensible realities we associate with the sublime can in any sense be objects of knowledge. Coleridge, on the other hand, is writing the autobiography of a literary artist. In doing so he makes very powerful claims for the close association of art and knowledge – specifically knowledge of supersensible realities. Consequently, he paid very little attention to *CJ*, and the attention that he did pay was only to those parts of it that he thought, mistakenly, could be identified with an Idealized interpretation of the treatment of the relation between noumenal and phenomenal reality in the first two *Critiques*. After 1811 probably, and certainly before embarking on the *Biographia Literaria* at some time before 1817, he read Kant in the spirit of Schelling's Idealization of his metaphysics, passing hurriedly over the parts of his philosophy – notably 'The Analytic of the Beautiful' – that didn't fit into Schelling's transformation of it, and ignored or distorted aspects of his system that he thought represented a threat to his ethical and religious intuitions. As a result, there is a fundamental discrepancy between Kant's and Coleridge's accounts of aesthetic experience, and this has had serious consequences for the development of post-Kantian theory in Anglo-Saxon literary circles. For Coleridge's transcendental interpretation of Kant's aesthetic, mediated first through Carlyle and then through later Idealist literary theorists, became the official version of it. As a result, the 'Kantian' foundation of the theory of art in England and America in the past two hundred years has born little relation to the aesthetic theory Kant actually invented in the book he devoted to that particular subject.

It is principally with the idea of the *noumenon*, or 'thing-in-itself', that Coleridge is preoccupied in his reflections on Kant. This is a notion Kant introduced in *CPR*, where it plays an important part in the demarcation of the scope and bounds of the faculties of reason and understanding. The status of the *noumenon* in Kant's epistemology is not at all easy to establish. It seems to mean: what knowledge of objects, as they are represented to the mind, is commonly supposed to be knowledge of. We can use the word in a positive or negative sense (according to the configuration of the rest of our theory of knowledge) to describe what lies beyond the knowledge (of phenomena) we acquire when the understanding brings intuitions under concepts. Interpreted negatively, it has something in common with Lacan's 'Real', 'before which the Imaginary falters, and over which the Symbolic stumbles'. But for Kant it is neither positive in the Coleridgean nor negative in the Lacanian senses. It is a limit, not a space, a means of defining what we can properly be said to know, not a guarantee that there is more that is knowable than our powers of understanding actually enable us to know.

In the succeeding *Critiques*, the concept of the *noumenon* makes possible an investigation of the moral and voluntary life which would otherwise have remained out of bounds to philosophy. But it doesn't do this by suddenly being represented as a possible object of knowledge – existing somewhere behind the external world of sense perception, or the internal psycho-theatre of the phenomenological or the transcendental ego. The one thing Kant makes unambiguously clear about the *noumenon* is that we can know nothing about it, that our understanding cannot encompass it. Though its possible existence is rationally demonstrable, its real presence is to be inferred only through the exercise of the will and the operations of practical reason.

Coleridge will not accept this interpretation of Kant's theory of knowledge, in spite of the fact that it is one of the cornerstones of the whole critical philosophy. Instead, he insists that Kant involved his argument about things in themselves in deliberate obscurity (they were among the 'few passages' in Kant that were obscure even to him) for reasons of prudence. Kant didn't want to go the way of his predecessor, Wolff, who was expelled from Prussia in 1723 for giving offence to his theological colleagues. Coleridge points out that Fichte's expulsion from Jena on a charge of atheism half a century later, in 1798, demonstrated that Kant had good reason to be cautious. So he implied symbolically meanings that he feared to communicate discursively. Coleridge writes:

> In spite, therefore, of his own declaration, I could never believe, that it was possible for him to have meant no more by his *noumenon* or THING IN ITSELF, than his mere words express; or that in his own conception he confined the whole *plastic* power to the forms of the intellect, leaving for the external cause, for the *materiale* of our sensations, a matter without form, which is doubtless inconceivable. I entertained doubts likewise, whether in his own

mind he had ever laid *all* the stress, which he appears to do, on the moral postulates.[29]

In spite of Coleridge's very definite ascription of a positive character to 'things in themselves' here, critics reviewing his work as a whole have expressed bafflement about whether he in fact discarded them from his own metaphysic. For Wellek, sometimes he doesn't (p. 81), sometimes he does (p. 95), sometimes he defends them as metaphors or symbols (p. 99). Coleridge's most recent biographer interprets his view of the status of the *Ding-an-sich* as a deviation from Kant's own more circumspect treatment of the subject: 'It was...characteristic of him to go a step further than Kant had gone and make a claim for our knowledge, through Reason, of universality and necessity, the very entities to which Hume had said we had no access, a position from which Kant did not dissent.'[30] Clearly what happened was that the notion of 'things in themselves' simply merged into a resplendent unity of mind and nature, in a manner entirely consistent with Coleridge's pre-existent beliefs and his more recent reading of Fichte, Schelling and other post-Kantian Idealists.

What Coleridge says Kant could not possibly have meant is precisely what Kant did mean. Indeed it was by meaning it that Kant was able to bring about his Copernican revolution in philosophy, removing the requirement that philosophers concern themselves with the existence or otherwise of worlds not given to us in experience but capable of becoming objects of knowledge. For Coleridge knowledge, and the language in which we communicate knowledge, must have a relation to what lies beyond the world as we experience it, and that relation must be a positive one. In this he is at one with his Idealist contemporaries and successors, especially with Hegel,[31] who conceived of art as an agency of knowledge every bit as effective as philosophical thought, which it shared with the life of the mind in general. It discovered in the apparent incoherence of experience a reality that is simultaneously ordered, spiritual and absolute. Like Schelling, Coleridge named the agency which made this discovery the Imagination (the secondary, esemplastic Imagination). In English, the most powerful celebration of this fittingness of the outer world of nature to the inner world of the human mind is in Wordsworth's 'Preface' to *The Excursion*:

...my voice proclaims
How exquisitely the individual mind

> (And the progressive powers perhaps no less
> Of the whole species) to the external World
> Is fitted: – and how exquisitely too –
> Theme this but little heard of among men –
> The external world is fitted to the Mind;
> And the creation (by no lower name
> Can it be called) which they with blended might
> Accomplish: – this is our high argument.
>
> (61–70)

By such means Coleridge sought to make use of Kant to open up extra-rational territory to the spirit, rather than to set limits to the territory we already categorically know.

This takes us back to the point Coleridge made in *The Principles of Genial Criticism* about the 'perceived harmony of the object... with the inborn and constitutive rules of the judgement and the imagination'. For Kant there can be no such simple equation of the faculty of judgement and the functions of the imagination. Nor does he assume there is an innate harmony between mind and nature of which the harmony between the different faculties of cognition are merely the mental reflections. There may come a point in Kant's philosophy of reason when some such harmony is argued for. But that has more to do with his inquiry into the sublime[32] and, later, into the principle of teleological judgement in the second part of *CJ*. Furthermore it is never viewed as a constitutive principle of judgement. In so far as it applies to objects in nature it does so to make possible a certain way of looking at nature – *as if* it were ordered, final, purposive. Where he is concerned with aesthetic judgements Kant is looking for a harmony of the faculties of mind created by the cooperation of the empirical understanding and the imagination quite different from that which takes place when judgements of empirical knowledge are being made. Coleridge, though, agreed with Schelling that the mind participates in a universal harmony which it reflects and which (because, ultimately, world and spirit are one) it includes. For both of them, the fundamental identity of mind and nature ceases to be, in Kant's terminology, a regulative idea of reason and becomes a constitutive principle of judgement. Thus for Coleridge to write that Kant's symbolic meaning is the opposite of his apparent meaning, and 'for those who would not pierce through this symbolic husk, his writings were not intended', was to misread Kant as a philosopher and recreate him as a Unitarian prophet. He more or less admitted as much in a letter to J. H. Green, in which he wrote that 'Reason is *subjective* Revelation, Revelation *objective* Reason...If I lose my faith in *Reason*, as the perpetual revelation, I lose my faith altogether.'[33] As a result, an inversion of Kant's epistemology became the basis of Coleridge's aesthetic, and Kant's actual aesthetic deliberations were largely ignored.

This is nowhere more apparent than in Kant's discussion of the unpurposeful purposiveness of art. The phrase Kant uses, '*Zweckmässigkeit ohne Zweck*', has been translated in widely different ways. Bernard translates it as 'formal purposiveness' or 'purposiveness without any representation of a purpose'; Meredith as 'the form of finality'; Warnock, literally, as 'finality without end'.[34] Abrams traces the concept to an essay by Karl Philipp Moritz five years earlier, where the phrase used is 'internal purposefulness' ['*innere Zweckmässigkeit*'].[35] There is no hint in Coleridge that he has ever read about this – another indication that he had more or less ignored what Kant had to say in the 'Analytic of the Beautiful'. The present argument demands that we turn our attention to what Kant actually wrote about purposiveness, or the Form of finality, in art, before moving to an examination of the importance of Coleridge's neglect of it. This will take some time, because it needs to be done against the background of Kant's general theory of art – in relation to which, and only in relation to which, his concept of purpose can be understood.

KANT'S *CRITIQUE OF JUDGEMENT*

> Literature...whose essence is power and not knowledge, was to him, at all parts of his life, an object of secret contempt.
> Thomas de Quincey on 'Kant in his Miscellaneous Essays'.

Kant's views on aesthetics are contained in a book, *The Critique of Judgement*, about the nature of teleological judgement. By this Kant means whatever faculty of the mind comprehends nature in its Form of finality, or purposiveness. Nature is conceived of as being adapted to mind in so far as its empirical laws fall under the principle of unity. Thus what lies outside the mind is conceived of as being in harmony with the mind, in so far as both submit to the principle of unity and purpose. The mind requires nature to be purposive in order that it shall appear to be intelligible, and so the purposiveness of nature becomes an a priori concept, or a transcendental principle of the faculty of judgement. The insistence that the principle is transcendental is important, because this means that although it is applied to actual or possible objects of empirical knowledge, it is not itself an object of empirical observation. This is why it is at the same time an a priori concept. If Coleridge had addressed himself to the subject of teleological judgement, as Kant expounds it in his third *Critique*, he would not have found any justification for assuming there was a principle of order in nature. The *Judgement* would have provided him with no excuse for repeating the error into which he had already fallen in his interpretation of *CPR*. Just as in *CPR* there is no indication that *noumenal* reality is to be understood as being more than a conceptual limit beyond the possibility of the existence of which cognition cannot stray, so in *CJ* there is no ground for supposing that pattern or design exists in nature as something separate from a requirement we make of nature in order that it shall appear intelligible to us. We need to view nature as if it were ordered. Whether or not it actually is ordered is beyond the powers of human intelligence to determine.

Students of Kant's philosophy have often been puzzled about the relationship between his treatment of the Form of finality (or purposiveness) in *CJ* and his treatment of the relation between phenomena and *noumena* in *CPR* and *CPrR*. They have also been puzzled about the way that, in *CJ* itself, Kant relates the principle of purposiveness in nature to our apprehension of ends in nature on the one hand, and in works of art on the other. Further, there seems to be a difference between the way our pleasure and displeasure in objects of aesthetic awareness functions in respect of what Kant calls, after the English writers of the mid-eighteenth century, the beautiful and the sublime. Any attempt to resolve these puzzles, or adjudicate between the views of those who espouse different interpretations of Kant's meaning, is beyond the competence and scope of the present argument.[36] This will touch on the connections between *CJ* and the two earlier *Critiques*, and between the two Parts ('The Critique of Aesthetic Judgement' and 'The Critique of Teleological Judgement') of the *CJ* itself, from time to time. But the main preoccupation will be with 'The Critique of Aesthetic Judgement' alone, as far as possible separate from its context in *CJ* or the wider context of Kant's general philosophy. Furthermore, within 'The Critique of Aesthetic Judgement' I propose to consider 'The Analytic of the

Beautiful' in isolation from the complementary 'Analytic of the Sublime'. This is because the bearing of 'The Analytic of the Beautiful' on the issue of literary intention is perceptible in a way that doesn't appear to be the case in 'The Analytic of the Sublime', and because 'The Analytic of the Sublime' has what I believe is a tendentious connection with Kant's treatment of the *noumenon* as a determinable reality, both in 'The Critique of Teleological Judgement' here, and in the discussion of the *noumenon* in the *CPrR*. In any case, many Kant scholars agree that the analysis of judgements concerning the sublime has no connection with analysis of judgements concerning the beautiful that is necessary to our understanding of the beautiful, even though the reverse might not be true. In a recent account of Kant's aesthetic, Mary McCloskey[37] opens her chapter on the sublime with the claim that 'Kant's analysis of judgements concerning the sublime cannot but make the impression of an interruption in the development of the main theme of the Critique of Aesthetic Judgement.'

In any event, most commentators have shied away from the second part of *CJ*, 'The Critique of Teleological Judgement'. Or, like Körner,[38] they have interpreted it in the light of Kant's interest in the biological sciences – a procedure which helps to make a limited amount of sense of ideas which Kant surely expected to have a more comprehensive application. Kant believed it was possible to use the idea of purposiveness in nature as a regulative way, at the same time as insisting that it has an application to nature in its objective character. If this were possible it would follow that it is also possible to give content to the notion at the heart of *CPrR* that there is a connection between the mechanistic world of the physical sciences and the world of freedom sensed in the background of our moral life and in our intuitions of the divine. One sees, then, why Kant must have attached so much importance to this part of the third *Critique*, since its function was primarily to bridge the gap between *CPR* and *CPrR* by introducing the concept of judgement as a separate faculty of the mind. This he set beside understanding and reason as a crucial third term in his psychological metaphysics.

Understanding makes possible empirical knowledge of natural science and mathematics by referring sensation to cognition in conformity with the logical categories Kant identified in the 'Analytic' of the *CPR*. Intermediate between sensation and cognition Kant places what he calls a 'schema': something given by the imagination that brings together an experience in sensation and the formal category applied to it in the act of cognition. It is important to understand what Kant means by this because it is the only activity recognized as an activity of the imagination in *CPR* and as such is a crucial link with *CJ*. Kant is trying to show how sense experience (he calls it 'the manifold of perception') falls under concepts in such a way as to be made over into an object of knowledge. He suggests it is logically possible to have the notion or concept of something, a table for example, and at the same time be incapable of applying it to specific instances of tables in such a way as to enable us to recognize them as tables. In order to do this we need to supply an intermediate entity, something defined as neither a concept nor a sense datum, that links a concept of a table with an experience of actual tables so as to fill out our formal understanding of tables in experiential terms. This is what a schema does. The imagination, then, operates on behalf of the understanding upon the stuff of experience, making possible, in association with cognition and sensation, empirical knowledge of objects.[39]

Unlike the understanding, the faculty of reason does not deliver empirical knowledge of anything. In *CPR* reason plays a largely negative role. Pure reason is the principal object of the critique of the title, and as such its claims to being a faculty of the mind at all are in doubt. If by reason we refer to a faculty of mind that delivers judgements about reality independent of experience, then we do indeed refer to something that does not exist. There can be no such judgements, because knowledge must be knowledge of what the understanding grasps in its engagement with experience. That is to say, we can give no content to propositions asserted by someone about ideas that have no relation to possible experience. Even so, we do entertain 'ideas' of reason, even if the functions they perform cannot be described as having a content. Such ideas are those of God, freedom and immortality – ideas which Kant describes in the 'Dialectic' as illusions of speculative metaphysics. They are illusions that Coleridge sought to pass off as cognitive realities, though in order to do so he too made a distinction between reason and understanding. According to Coleridge, Kant wanted to say that reason was the agency of a higher form of cognition, which it then transpired it was the business of the imagination to express through art and literature. Really, though, for Kant such ideas of reason have only a regulative function. They make it possible for us to think of the world in general terms, as if it were explicable by reference to the findings of speculative metaphysics. However, all this changes when

Kant shifts his attention from knowledge of the world and nature to knowledge of our moral consciousness. Here reason in its practical character, reason engaged in determining what we ought to do, presupposes all those ideas that had a merely regulative role in the activities of pure reason. It is not at all clear to what extent Kant sees the practical reason as having a constitutive or a regulative function. It appears that the fact of moral choice confers reality in terms of will, or desire, on what was merely illusory in terms of cognition, or understanding. Pure reason masquerades as understanding when it presumes to give a content, in the form of objective knowledge, to the findings of speculative metaphysics. But practical reason appears to give a content to those same findings by making them conform to the will. From the point of view of Kant's aesthetics we need to bear in mind that, while he certainly argues that the ideas of speculative metaphysics are illusory in relation to pure reason, those same ideas have a kind of content in relation to practical reason, and practical reason is what Kant was concerned with when he wrote about the mediating role of the judgement between understanding and reason in *CJ*.

Kant's picture of the mind can be viewed as follows. Sensibility is always present as a capacity of the mind to respond to a stimulus. The faculty of understanding makes possible perceptual knowledge by being brought to bear on objects of sensibility through the application of concepts and the intercalation of schema produced by the imagination. The faculty of reason, in its pure identity, seeks to deliver concepts without reference to experience, i.e. without the collaboration of sensibility, and therefore without the raw material of sense data. As a result, what it actually delivers is metaphysical illusion. In its practical identity, though, it legislates for desire. That is how, although it does not make possible perceptual knowledge, it does serve to legitimize those acts of will and conscience that constitute our moral and religious life.

Mediating between understanding and practical reason is the faculty of judgement. It does this in two ways, which Kant calls determinant and reflective. Determinant judgement, which doesn't concern us here, applies to universal laws, rules or principles, discovering the particular instances that are to be subsumed under them. By contrast reflective judgement, judgement defined as a separate faculty from understanding and reason, applies to particulars. Moving beyond consideration of particulars, it will do one of two things. Either it will repeat the activity of the determinant judgement, but inversely, seeking to discover general laws that are to be inferred from particular instances. Or it will refer back the effect of its contemplation of particulars to the system of mental faculties of which it is a part, revealing as it does so a sort of fittingness between the mind and its objects. I use the word 'fittingness', which is not Kant's, to avoid premature and finicky distinctions between a form or structure on the one hand and an activity on the other. Clearly it is possible to describe mind and the intentional objects of mind in terms of either or both of these things, as I shall do below.

Reflective judgement is related to pleasure because it legislates for feeling, in much the same way as, a priori, understanding legislates for empirical perceptual knowledge and reason legislates for desire. Feelings of pleasure and displeasure are therefore closely associated with the a priori principle of judgement, not least where judgement is exercised in respect of objects of aesthetic contemplation. Since the a priori principle of judgement has to do with the consideration of ends, or finality, or purposiveness, descriptions of aesthetic contemplation will tend to place special emphasis on how works of art give pleasure by working on feelings that bring such considerations to mind. The ability of the mind to sense its own workings as of their nature end-directed, or characterized by the Form of finality, is at the same time for it to experience feelings of pleasure. Kant takes it for granted that pleasure is strongly associated with a sense of order and harmony, and he shows that this is inconceivable apart from notions of finality or purposiveness. This is what it means for judgement to mediate between understanding and practical reason. However, when it comes to explaining how feeling mediates between these other two faculties, commentators on *CJ* entertain very different opinions. Roughly speaking, these have to do with whether, in matters of taste (as Kant calls aesthetic judgements), the reflective judgement functions mainly in association with the understanding or with practical reason. In 'The Critique of Aesthetic Judgement' it seems to me that he emphasizes the link between judgement and understanding, i.e. between the faculties of feeling and cognition. It is true that the connection with practical reason gives a practical role to the ideas of speculative metaphysics in the exercise of reflective judgement. But that is not really where Kant places the emphasis in the first part of *CJ*, and even if it were, it would not stand in any interesting or useful relation to those ideas of intention with which the present argument is concerned. On the other hand the connection with understanding is crucial because of the role the imagination plays in it. It is to the way imagination facilitates the cooperation of feeling and cognition that I now turn.

In matters of taste, understanding cannot be brought into play in the same way as it can be, and is, in matters of empirical judgement, because feeling does not provide us with knowledge of anything. Yet feeling cannot exist apart from objects of feeling, and objects of feeling must in some sense be known to the person who feels them. So the question arises: is the connection between aesthetic experience and knowledge merely a matter of temporal and causal sequence? Does the understanding bring the objects of sensation under concepts, thus making knowledge possible, and then knowledge deliver its objects to the faculty of feeling, which converts them into the proper material for aesthetic judgement? Apart from the fact that this sounds far too mechanical to be plausible as a description of the psychology of aesthetic judgement, it contains two flaws, of omission and commission. It ignores what Kant wrote about the imagination's role in making intelligible to experience the bringing of sensations under concepts, and it assumes that thinking in general, not just cognitive thinking, is inseparable from the use of concepts. To grasp Kant's idea about the way the imagination works, it is first necessary to grasp what he thought about how concepts are used in aesthetic judgement.

Kant's analysis of the beautiful is broken down into four distinct but related moments (as he calls them; we would be more likely to call them something like 'necessary elements') of taste. Two of these, the second and the fourth, stipulate that what is being defined is to be understood as being 'apart from a concept'. The second moment reads: 'The beautiful is that which apart from concepts is represented as the object of a universal satisfaction.' The fourth reads: 'The beautiful is that which without any concept is cognized as the object of a necessary satisfaction.'[40] What does Kant mean here by 'apart from a concept'?

It isn't difficult to see what he means if we are thinking about what judgements of taste are actually like. Characteristically they take the form of: 'I feel such and such about this poem, painting, piece of music, and I think you should feel this way as well.' In other words it is an inter-subjective judgement, starting from description of a private response, and moving on to a recommendation that this response should be shared by all who read, see or hear the same work of art. Of course the categorical recommendation presupposes that all that is merely personal in the immediate response to the work has been abstracted from the response as it is now articulated. Indeed this is what is required by the first moment of taste, which states that 'Taste is the faculty of judging of an object or a method of representing it by an entirely disinterested satisfaction or dissatisfaction.'[41] This has been variously interpreted, but everyone agrees that by 'disinterested' Kant means that idiosyncrasies of response have no role to play in ascriptions of beauty to whatever falls under judgement. This statement contains the assumption that it is possible to separate out the accidental properties of private interest and the necessary and universal properties that exist in some sense apart from it. That is what the definitions of the second and fourth moments of taste insist on. Furthermore this separating-out process can be achieved by each individual without reference to some pre-existent idea or concept of what should be the case. There is no definition of what constitutes the beautiful that can usefully be brought into association with particular instances offered to taste as candidates for judgement. None of Kant's four moments can be interpreted as if it were part of such a definition. Instead, each of them is to be understood as a description of what people are in fact doing when they give expression to judgements of taste.

Kant calls the offering of a particular work of art or aspect of nature as a candidate for judgement a 'presentation'. When a presentation takes place, the mind becomes conscious of an interplay of the faculties of understanding, feeling and reason, which pleases because something in the presentation satisfies the mind's innate requirement of order, harmony, finality. What that 'something' is, Kant doesn't inquire into. After all, he is not a critic. He is more interested in describing the aesthetic response on the other side of the presentation, by defining beauty, in the third moment of taste, as 'the form of the purposiveness of an object, so far as this is perceived in it without any representation of a purpose'.[42] This is the central feature of Kant's aesthetic, and entails the vital connection between purposiveness and intention that I want to consider below. But for the moment the emphasis must fall on the reference of whatever it is in the object that appeals to the mind as a perceiving subject, and the association of that with a categorically formal value. Thus we can picture the mind in the act of aesthetic judgement disinterestedly responding to a presentation by referring it formally and non-conceptually to feeling. In doing this it cooperates with the mind's predisposition to find satisfaction in general notions of purposiveness and finality that have some sort of deep connection with, or in some way reflect, the mind's own workings and constitution. As Richard Wollheim says, 'All rests on the fact that deep feelings pattern themselves in a coherent way all over our life and behaviour.'[43] Kant's is one very convincing way of showing how this happens from the receiving end, as it were, of the artistic process.

If aesthetic judgement works through feeling by detaching understanding from concepts, what part is played in it by the imagination, which, as we have seen, attaches concepts to sensations in order to make them intelligible to the understanding as empirical knowledge? In *CPR* Kant argued that the role of the imagination in the psychology of understanding is merely instrumental. It is there to preside over the schematizing of concepts, and since concepts are governed by rules, this means that the exercise of the imagination is constrained by those same rules. Kant takes the example of a dog. The concept of a dog includes such specifications as that it is an animal with four legs, carnivorous, includes several different breeds, is often kept as a pet or companion, *etc*. These are all formal attributes and without the intervention of any other mental faculty would come within the scope of the understanding merely as such. There is no way in which the intuited sense data could be brought under cognition so as to be recognizable as a dog rather than a barrage of sensations to which certain formal descriptions are to be attributed. How then do we recognize a dog as a dog? Not, Kant explains, by specifying yet more formal rules to close the gap between rule-bound concepts and unregulated sensations. And obviously not by enlarging or intensifying the power of sensation itself. (This is impossible because sensation is a passive capacity, not a directed mental power.) The faculties of intuition and understanding don't, as it were, have edges that stick together. There has to be something else that will act as a sort of glue at the metaphorical interface. This is what the imagination does. It has been described as filling out the formal understanding in experiential terms. In *CPR* that is all it does. Open up the gap again between intuition and understanding, and it just trickles away.

In *CJ*, however, Kant has decided otherwise. When the mind is engaged in the process described in the four moments of taste – disinterested, feeling a necessary and universal pleasure, and experiencing a sense of purposiveness without purpose – the imagination is freed from concepts and acquires an autonomous function it never possessed in the earlier *Critiques*. It is still not autonomous in the sense of being able to flourish apart from the understanding altogether. But it is autonomous in so far as it is no longer in the service of the understanding. The aesthetic judgement can be brought into play only on condition that the imagination is allowed to cast off its merely instrumental role in making empirical knowledge available to the understanding. It has acquired a function of its own which it can perform in association with, but not in dependence on, the understanding. It can do this because aesthetic judgement has nothing to do with concepts. Or, rather, nothing to do with concepts as they have been defined in relation to other kinds of judgement. 'Only where the imagination in its freedom awakens the understanding, and is put by it into regular play, without the aid of concepts does the representation communicate itself, not as a thought, but as an internal feeling of a purposive state of the mind.'[44]

Kant's antinomy of taste suggests that aesthetic judgement cannot be both aesthetic and a judgement because it cannot be both an expression of subjective experience and the statement of an objectively valid proposition. He solves the antinomy by showing how the exercise of the imagination in aesthetic judgements involves only indeterminate concepts – ideas of order, pattern, regularity – that are not attached to particular objects recognizable as such. In fact it is unlikely that Kant believes this is the only way the imagination can be conceived of as working in free association with the understanding, but his lack of interest in the works of art that make presentations possible sometimes makes it seem as if he does. Some commentators on *CJ* suggest that in the free play of imagination, concepts might well be determinate but not applied. A concept might enter into the way imagination and understanding cooperate in making sense of a sensation, but it will be applied to what the object of sensation is taken *as*, rather than to the object itself. In aesthetic judgement a representation is always experienced as a representation, not as whatever it is that is being represented. Roger Scruton gives the act of seeing a set of marks as a pattern as an example of the indeterminate free play of the imagination, and that of seeing a picture of a face as an example of the determinate but unapplied free play of the imagination.[45] Literary and musical analogies will readily suggest themselves.

What needs to be done now is to refer these indeterminate functions of the imagination back to the wider context of mental activity to which, in the act of aesthetic judgement, they belong. This can only be done by taking more thoroughly into account Kant's views, in the third moment of taste, about the relation between harmony and purposiveness. It is in that relationship we discover the formal unity the mind attributes to objects and objects reflect back upon the mind.

First we need to remind ourselves of what Kant said about the form of purposiveness in the third moment of taste. 'Beauty', he wrote, 'is the form of the purposiveness of an object so far as this is perceived in it without any representation of a purpose.' It matters, though, what sort of object we have in mind. There is a difference between a feature of the natural world on the one hand and a work of art on the other – where both are viewed as candidates for aesthetic judgement. I have indicated that for obvious reasons this book does not concern itself with the first of these objects in its own right. Nevertheless such objects are important in relation to works of art, which share some of their characteristics (again, as candidates for aesthetic judgement) but don't share others. What they share is summarized by McCloskey in the opening of her chapter on fine art. She points out that, for Kant, both successful works of art and beautiful natural objects

> are to be judged universally and necessarily pleasing [according to the second and fourth moments of taste] in virtue of their being of a perceptual form which is suitable for setting the cognitive powers, imagination and understanding, into harmonious free play. Such universality and necessity is, for both kinds of examples, underpinned by the finality [according to the third moment of taste] of such forms for human perception.[46]

McCloskey shows that according to Kant the two main points of difference between successful works of art and beautiful natural objects are that successful works of art are intended to be as they are, and that they are expressive of 'aesthetic ideas'. She then – quite rightly, because Kant makes much of this too – devotes a great deal of space to consideration of aesthetic ideas, separate from the intentional character of the work of art that in some sense refers to them. But aesthetic ideas conceived of in this way are something of a distraction. It is possible to include what they are supposed to contribute to a work of art with reference to nothing more than the intentional character to which Kant in any case devotes a considerable amount of thought.

Works of art differ from features of the natural world as objects of aesthetic enjoyment by virtue of their intentional aspect. If it transpires that we can either forget about aesthetic ideas or subsume them under the notions of intention, as these will be defined below, then they will be in all other respects the same. This is a very large step to take, and arguments advanced by Wollheim and Goodman[47] against the assumption that there is anything like this degree of similarity between judgements about, say, a tulip, and judgements about, say, a poem about a tulip, are cogent. But this is because of the altogether too great weight given to the presence of aesthetic ideas in art but not in nature, and the failure to give as full a weight of emphasis to the notion of intention as it deserves. I shall argue that in a sense the way intention functions in a work of art *is* the aesthetic idea that the work of art embodies. I shall also argue that the purposiveness without purpose Kant insists upon in his third moment of taste is a feature of natural beauty only by virtue of the way, inevitably, we transfer what we expect to find in works of art to what we have grown used to expecting to find in the natural world. Kant writes:

> In a product of beautiful art, we must become conscious that it is art and not nature; but yet the purposiveness of its form must seem to be as free from all constraint of arbitrary rules as if it were a product of mere nature. On this feeling of freedom in the play of our cognitive faculties, which must at the same time be purposive, rests that pleasure which alone is universally communicable, without being based on concepts. Nature is beautiful because it looks like art, and art can only be called beautiful if we are conscious of it as art while yet it looks like nature.[48]

To common sense it often seems as if it should be the other way about,[49] and that we artificially transfer to our appreciation of works of art (especially visual works of art) what 'comes naturally' to us in our appreciation of the world of nature. Kant explains why this is a mistake by laying proper emphasis on intention in aesthetic judgements of works of art as a constitutive idea (i.e. certain perceptual forms really are final for perception), and intention in aesthetic judgements of nature as a regulative idea (i.e. certain perceptual forms are final only in a formal sense). It remains a fact that it is easier for us, intuitively, to appreciate Kant's point about purposiveness, or finality, by thinking about the beauty of natural objects than about the beauty of works of art, and that is what I shall do below. The aim will be to make the point by reference to aesthetic judgements about natural objects, then to transfer the reference by analogy to works of art in such a way as to modify the appropriateness of that earlier reference. This will be done by accounting for the role of intention, or the idea of intention, in each case.

Perhaps the clearest explanation of Kant's third moment of taste is in Copleston's reflections on the beauty of a rose:

> If we look at a flower, say a rose, we may have the feeling that it is, as we say, just right; we may have the feeling that its form embodies or fulfils a purpose. At the same time we do not represent to ourselves any purpose which is achieved in the rose. It is not merely that if someone asked us what purpose was embodied in the rose we should be unable to give any clear account of it: we do not conceive or represent to ourselves any purpose at all. And yet in some sense we feel, without concepts, that a purpose is embodied in the flower. The matter might perhaps be expressed in this way. There is a sense of meaning; but there is no conceptual representation of what is meant. There is an awareness of consciousness of finality; but there is no concept of an end which is achieved.[50]

The link between feeling and non-conceptual functions of understanding has already been discussed. Now we have to attend to the link Kant is seeking to establish between feeling without concepts and feeling that something we would usually describe with reference to concepts (of purpose) is a necessary formal constituent of any object of reflective judgement. It is significant that Kant himself often takes a flower as an example of such an object,[51] and that Copleston follows his lead here. Both profess to see a process of unfolding, the unfolding of a purpose, in the form of a flower, and many other writers on this subject have done the same. Here, for example, is G. M. Hopkins describing in his *Journal* (13 June 1871) what a flag flower looks like:

A beautiful instance of inscape sided on a slide, that is successive sidings of one inscape, is seen in the behaviour of the flag flower from the short bud to the full blowing: each term you can distinguish is beautiful in itself and of course if the whole 'behaviour' were gathered up and so stalled it would have a beauty of all the high degree.[52]

And here is Wittgenstein on spring blossom:

One might say: art shows us the miracles of nature. It is based on the concept of the miracles of nature. (The blossom, just opening out. What is miraculous about it?) We say: 'Just look at it opening out.'[53]

These passages are not quoted in order to enlist Hopkins or Wittgenstein as subscribers to Kant's aesthetics in general. But the choice of words to describe the beauty of the flower or the blossom is suggestive, and Wittgenstein's implied distinction – between awareness of concepts and an articulated response based on concepts but unaware of them in the immediacy of feeling – is basically Kantian.

Looking at a rose, we feel that it is 'just right'. But what are we looking at? At the petals folded one on the other forming the core of the unopened flower, or spread out in full bloom exposing the corolla to view? At the way the coloured base of the petals rises out of the green calyx at the top of the stem, or the pattern of colour repeated with variations of tone and hue from one open petal to another? Or the relationship between leaf and flower in respect of shape, colour, the impression they severally give of softness and sharpness, light and dark, clear or wavering definition? Copleston's description is too abstract and exemplary to be of much use here. In both Hopkins and Wittgenstein it is noticeable that the emphasis falls on activity wrought into an intensely realized shape. Wittgenstein's 'Just look at it opening out' allows the stress to fall simultaneously on the object, 'it', and the activity, 'opening out'. Hopkins' flower is characterized by its 'behaviour' or 'succession of aspects' 'stalled' or 'gathered up' in a single unity of what he calls 'terms'. This is like Dean Jocelin's vision of the apple tree being 'more than one branch' at the end of Golding's *The Spire*; or like Yeats's chestnut tree, in 'Among School Children', comprising the root, leaf, blossom and bole – both of them instances of a single process wrought into unity. In Yeats the activity is compared with the artistic unity of the dance, created out of the several movements, swayings, glancings of the dancer. So it is not a single object or attribute, but a process we derive from those relationships we can only describe to ourselves in terms of purpose. There must be a point to it, a requirement that we speak in terms of fulfilment, achievement (even Robert Graves's wax florist rose makes claims), purposiveness.

But without purpose. The tulip, rose, flag flower, blossom seems to embody a sense of purpose without having any particular purpose that could be described by reference to concepts. It has a form that is final for perception in the sense that we feel in it an end-directedness. Without having any notion of a particular end to which it is tending, we nonetheless feel that there needs to be included in our response to it a sense of its being *for* something, of its having an end which is in some way internal, having no application to further ends that lie outside itself. This was what Kant meant by claiming that aesthetic judgements are subjective, formal and internal. However, when we move from aesthetic judgements of natural objects, like roses, to aesthetic judgements of works of art, like poems about roses, we find that the subjective character of the judgement feels more pressingly related to the subjective character of the judgements of others. And the more this is felt, the more the sense of purposiveness within the object seems to approximate to an actual purpose that can be explained in conceptual terms. Of course it never becomes identified with the sense of a particular purpose. If it did, as Collingwood argued, it would be the product of a craft, not of fine art. In any case Kant is categorical on the separation of the understanding from concepts in aesthetic judgement. Nevertheless the distinction between aesthetic judgements of nature and aesthetic judgements of works of art can be understood only by recourse to a parallel distinction between different degrees of approximation to a sense of particular purpose within the general scope of intransitive purposiveness that contains them both.

At the back of this distinction lies the notion of intention that plays an important role in our judgements of works of art, but not in our judgements of the beauties of nature. In Kant, though, this fact is disguised by the faculty psychology he uses to describe the mind in the act of making aesthetic judgements about works of art, and by his theory of aesthetic ideas that he believes contribute so much to the making of those judgements.

It is not immediately apparent that a reflective judgement of indeterminate formal purposiveness in a beautiful work of art must have something to do with the recreation of the intentional activity of one person's mind in the intentionally active mind of another. I want to argue, though, that there is a necessary connection between these two things, that they are two different ways of describing the same activity. It is to be hoped there is some such connection because, otherwise, like Kant, we would have to import into our aesthetic theory the notion of aesthetic ideas that played such an important role in laying open the most valuable part of his theory to the misunderstanding referred to at the opening of this chapter. Aesthetic ideas, with all the lumber of notions of the sublime and appeals to practical reason that inevitably accompany them, proved to be the most acceptable parts of the theory for writers of an Idealist tendency (like Coleridge). But as well as being difficult to justify, even in the terms of Kant's own psychological model, they are also surplus to requirement if we look at the rest of the theory in the way I am proposing we should. For Kant they were not surplus to requirement, because without them he could not define the indeterminate concept that underlies the judgement of taste in such a way as to bring into close proximity the bases of aesthetic and moral judgement. But we don't need to labour under the same constraints. We are under no obligation to fit the valuable parts of Kant's aesthetic into the wider metaphysical scheme of the three *Critiques*. All we need to do is to see if the basic argument of 'The Analytic of the Beautiful' is coherent and usable when the terms of contemporary philosophers of action and intention are substituted for those of the faculty psychology Kant himself had used.

The aesthetic ideas of 'The Analytic of the Sublime' get in the way of our doing this because they introduce into Kant's theory of aesthetic judgement two things that don't need to be there. These are the intervention of the faculty of reason, and a substitute for concepts in the form of the sensible counterpart of a rational idea.[54] In spite of his formalist approach to art, Kant could not rid himself of the idea that the objects of sensible intuition present in it – the characters in a play, the landscape in a picture, the images in a poem – must have a value in and for themselves, over and above the design the reader or viewer abstracts from the relations observed between them. Crawford quotes a passage from *CJ* (I. 2. 49) in which Kant approves of the way the poet 'ventures to realize to sense, rational ideas of invisible beings, the kingdom of the blest, hell, eternity, creation *etc'*. Even if he describes things of which there are examples in experience – like death, envy and fame – he uses his imagination, we are told, 'to go beyond the limits of experience and to present them to Sense with a completeness of which there is no example in nature'. Scruton[55] might have had this passage in mind when he supplied the example of Milton (of whom Kant surely was thinking: *Paradise Lost* was among the poems he most admired) expressing the vengeful feelings of Satan: 'We feel that we are listening not to this or that, as one might say, "contingent" emotion, but to the very essence of revenge. We seem to transcend the limitations contained in every natural example and to be made aware of something indescribable which they palely reflect.' Not the least important phrase Kant uses here, though, is 'ventures to realize'. What we are responding to in *Paradise Lost* is not Milton's achievement in making rational ideas available to sense, but his aspiration so to do. Kant had written (in the passage quoted above) that a rational idea is 'a concept to which no intuition (or representation of the Imagination) can be adequate', so even Milton could not produce the images that would in fact represent such an idea. Mary Warnock speaks for poetry as well as philosophy when she claims that 'In the *CPR* the ideas of reason are introduced in an almost entirely negative way. They stand for what we cannot conceptualize, and thus for what we cannot in any way experience in the world.'[56]

However, Kant has offered hostages to fortune by describing aesthetic ideas in these terms. The border between aspiration and achievement is so blurred that, at the level of ambition on which Milton is working in *Paradise Lost*, they appear virtually indistinguishable. It is important to remember also that the rational ideas which are 'almost' in our sensible experience of poems, paintings and music include a great deal more than invisible beings and abstract virtues and vices. They include, for instance, such cognitively incomprehensible notions as God, freedom and immortality. So the intervention of reason, operating on the 'genius' of the poet expressing itself through what Kant calls '*Geist*', or 'spirit', clears the way for something like the sensible apprehension of what, in *CPR* and *CPrR*, was exclusively in the domain of the *noumenon*. This was Kant's ulterior motive for introducing this unnecessary complication into his theory of aesthetic judgement. In spite of the fact that he doesn't use the Kantian term 'aesthetic ideas', it isn't difficult to see that it is this part of 'The Analytic of the Sublime' that stuck in

Coleridge's mind, linking up with what he had read about 'things in themselves' in the first two *Critiques*.

Coleridge ignored Kant's deliberations on the third moment of taste, preferring to develop his arguments about aesthetic ideas. But these make sense only in the light of Kant's wish to forge a link between reflective judgements of intermediate formal purposiveness and the *noumenal* world lying outside the scope of everything but the faculty of practical reason. If an alternative explanation can be found for Milton's ability to create for his readers the illusion that they have been placed in contact with supersensible realities (as I suspect it can), then we can argue that it was unnecessary for Kant to introduce this enormous distraction into his theory of art.[57] Coleridge isolated it from the rest of 'The Critique of Aesthetic Judgement', because what he was really interested in was ideas of the sublime. In accordance with Schelling's transformation of the whole critical philosophy, ideas of the sublime could be assimilated to his metaphysic of the Ideal coincidence of mind and nature and to his poetic psychology of the secondary Imagination. But in his description of aesthetic ideas in 'The Analytic of the Sublime' Kant had already provided some of the terms of this transformation. Here we find the imagination 'going beyond the limits of experience' which 'cannot be completely encompassed and made intelligible by language' – at any rate, one surmises, the language of 'Dejection: An Ode' as distinct from that of 'Kubla Khan'. Aesthetic ideas 'arouse more thought than can be expressed in a concept determined by words'. They have access to 'more than could be comprehended in a concept and therefore in a definite form of words' (*CJ*, I. 2. 49, pp. 157–9). However, we are at liberty to do the opposite of what Coleridge did, and confine our attention to Kant's notion of the indeterminate formal purposiveness of works of art in 'The Analytic of the Beautiful', separate from all considerations of the rational aesthetic ideas that intrude into 'The Analytic of the Sublime'. When we have done this, how can we tease out the implications of the phrase in terms that would be comprehensible to a contemporary philosopher of language and meaning?

The phrase is 'indeterminate formal purposiveness', or 'the Form of finality in an object, so far as it is perceived apart from the representation of an end'. The point I wish to make about descriptions of intention is that there is as much difference between what is understood by either of these phrases and determinate purposes with a particular content as there is between intentional acts viewed as sequences of intentions in doing and intentional acts (from a retrospective position, the same intentional acts) viewed as sequences of intentions of doing, or intentions to do.

Crawford makes clear the connection between concepts, purposes and intentions in Kant by referring to a passage from 'The Analytic of Teleological Judgement'.[58] The received view is that an object has a purpose when it is judged to be the result of an application of a concept, which can only be supplied by a person. 'Thus purposes (aims, goals, intentions) are linked to wills,' i.e. the workings of the faculty of desire through concepts. Then Kant moves beyond the received view by adding that we call an object purposive even when we cannot attribute its form and organization to a purpose that is the expression of a particular act of will. The fact of formal organization alone, inevitably accompanied in our minds by the idea of a concept (again, not any particular concept) lying behind it, is sufficient to communicate a sense of purposiveness of the object.

This is where the example from 'The Analytic of Teleological Judgement' might help to clarify the issue:

> Suppose a man in an apparently uninhabited land perceived a geometrical figure, say a regular hexagon, inscribed on the sand. His reflection, working on such a concept, would attribute, though mysteriously, the unity of the principle, of its genesis to understanding, and consequently would not regard the sand, the neighbouring sea, the winds, or beasts with familiar footprints, or any other unreasoning cause, as the ground of the possibility of such a shape. For the chance of not encountering such an object would seem so infinitely great that it would be just as if there were no natural law, no cause in the mere mechanical working of nature capable of producing it; but it would be as if only the concept of such an object, as a concept which understanding alone can supply and with which it can compare the thing, could contain the causality of such an effect. The figure, then, would be regarded as a purpose, but as the product of *art*, not as a natural purpose.

The hexagon functions like the watch at the opening of Paley's *Natural Theology*.[59] The presence in an object of something recognized as design or pattern brings with it the requirement that it be referred to a concept. This in turn implies the presence of a purpose embodied in the formal disposition of the object. The purposive condition of the figure is to be regarded as a product of art rather than of nature because of the element of human intention we inevitably infer in it. But the inference of human intention does not, in this instance, include the presence of some power of will. The concept, therefore, is not governed by the faculty of desire. Its relation to the object is less determinate than would have been the case, for example, if what the man had perceived was a tool. On the other hand if it had been a piece of rock (Paley had offered a stone) weathered into an interesting shape, there would have been no need for any reference to a concept at all. The hexagon occupies an intermediate position between tool and rock because of the way in which purposiveness necessarily inheres in it (unlike the rock), while the reference to a concept, which automatically accompanies a sense of purposiveness, is indeterminate (as contrasted with the tool).

Kant's choice of the hexagon, with its mathematical structure and its appearance on a stretch of uninhabited land, is typical of his approach to works of art. It is demonstrably regular (we don't have to argue for the fact of design) and it lacks a context. In these respects it is unlike works of art, especially literary ones, which rarely possess anything approximating to a mathematically regular form and which arise out of special circumstances of periods, genre, social organization, etc. These are unignorable features of poems and plays that might become candidates for our appreciation. So is the fact that their author was a person. Not this or that particular person, possessing particular and idiosyncratic traits of character, but a person possessing those intentional powers that are necessary qualifications for being any sort of person at all. Unlike Kant's hexagon, *Hamlet* or *Othello* or *King Lear* were manifestly created by an identifiable person in circumstances that were culturally and historically distinct. And unlike 'a man' in Kant's illustration, we respond to Shakespeare's plays in the special circumstances of our historical period and cultural environment. These facts complicate the notion of formal purposiveness that is being drawn to our attention when we substitute the one activity for the other.

Mainly, it does this by approximating to a determinate sense of purpose, without ever becoming identified with it. I have already suggested that this accounts for the principal differences between aesthetic judgements of natural objects and aesthetic judgements of works of art, and I am now adding to that suggestion the view that when the works of art in question are literary, the approximation tends to be an especially close one. Kant himself seems inclined to discriminate not merely between aesthetic judgements of natural and artificial objects but also between aesthetic judgements of artificial objects of greater and lesser density of conceptual definition. But when he does this he seems to have in mind not so much the relation between an achieved work of art and a preconception in the artist's or spectator's mind of what that work of art will be, as the relation between the materials of artistic expression conceived as representing nothing but themselves and the disposition of those same materials conceived as representing forms that can or must be described by reference to concepts. There are some forms of artistic expression that cannot be conceived of as representing anything but themselves. Instrumental music and abstract designs (wallpapers, for instance)[60] are offered as examples of such non-representational art, which Kant describes as being subject to judgements of free, as contrasted with dependent, beauty. Literary works of art must, though, be examples of dependent beauty, because they are made out of words, and words must refer us to concepts. But to say that the individual properties of a work of art happen to refer to concepts is something very different from saying that the aggregate of those properties in the form of the work of art as a whole does anything like the same thing. The formal relations that hold between the several conceptual units (the words) of a work of literature stand in the same relation to purposiveness as the relations between each individual unit and its semantic meaning stand in relation to purpose. The connection between intention and purpose, or purposiveness (as the case may be), is accordingly more or less close in proportion to the distance each different sequence of words travels from the governing concept that gave substance to the prior intention of expressing something in the first place.

When words are uttered separately, we have no verbal context which serves to bestow on them a precise definition. Dictionaries try to provide them with one. But when we use a dictionary, we are already trying to locate one among a range of meanings provisionally attributed to a word that has already been found in another context. In a psychological monograph that has much to say about participation in and transcription of conversations, Vygotsky provides a necessary qualification: 'A word in a context means both more and less than the same word in isolation: more, because it acquires a new context; less, because its meaning is limited and narrowed by the context.'[61] Not to know what a word means at all is not to read it as a blank, apart from its sound values, but to attribute to it the widest possible range of alternative meanings permitted by its grammatical position in the sentence. To know exactly what it means (if that is possible – on reading it for the first time, or coming back to it after having consulted a dictionary) is still to define it against a background of concepts that provide it with some sort of intellectual context. The context grows narrower and more restrictive as word is added to word in a developing sequence of structural relations which are inevitably and concurrently accompanied by semantic, i.e. conceptual, relations too. At the same time as the spoken or written utterance is being completed, the spacious context of possible meanings the listener brings to it is first infiltrated, then occupied, then eliminated. It looks as if the realization of the speaker's intention is achieved at the expense of the progressive shrinkage of the listener's expectations within the bounds of that intention.

If this were in fact what happens, communication between human beings would be more efficient and less interesting than it actually is. There are occasions when we do aspire to some such model of perfect linguistic understanding in our dealings with one another. When conducting business on the telephone, writing about constitutional reform to *The Times*, telling our solicitor what we want to put in our wills, we hope to achieve as complete a takeover of the broker's, editor's, solicitor's permissive linguistic mental context as is humanly possible. Even in less practically end-directed circumstances – for example, parents discussing their children, talk among families at the dinner table – we value the efficient communication sketched out above. Contributing parties to both types of communication probably act in accordance with maxims (of informativeness, truth, relevance and perspicuity) relating to the four subcategories of Grice's Cooperative Principle. The main difference lies in the way each party to the proceedings expects the other to bear more or less equal responsibility for the continuation of the exchange of views. There is an element of deliberation in the way conversational space is left unoccupied and the contextual perimeter of the discussion imprecisely defined.[62] It remains the case that immediately after these less formal occasions, the people who have participated in them will go away with the feeling that more has been understood than has been left pending for the next one;

though if nothing were to have been left pending for the next one, what had been happening would approximate too closely to the business conversation that in other respects is so much like it.[63] But literary art is not like it at all. It is at the furthest remove from the business-conversation end of the spectrum of verbal communication, because of the relation that obtains in it between the intentional activity of the writer and the supposed and actual expectations of the audience for whom he is writing.

Kant implies something of the same sort through his use of faculty psychology. He emphasizes the disinterestedness of aesthetic judgement by divorcing understanding from the application of concepts in two ways. One of them, the mediation between understanding and practical reason through the agency of aesthetic ideas, I have chosen to ignore. The other, the free play of the understanding and imagination through their application to the formal aspects of a presentation, is much more susceptible to reinterpretation and restatement in terms of the intentional vocabulary I have used in my commentary on Shakespearean tragedy. Talking about painting or music we might have to use a different vocabulary, or at least make significant modifications to the vocabulary I am using here. But literary art, with its building blocks of intentionally, because conceptually, charged words, displays precisely that combination of purposiveness and lack of determinate purpose that is the subject of Kant's third moment of taste.

The purposiveness exists at both ends of the writing activity. It is present in the intentional material out of which the separate phonemes, words and phrases of a poem are constructed. And it arises out of an intention of doing that might be adumbrated in a specific statement from the text (see Milton below, pp. 172ff); or in a more general notion of what is being undertaken, such as is being represented in what we assume to have been Shakespeare's decision to write another version of the story, or even the play, of, say, *Hamlet* or *King Lear*; or by something less tangible that we intuit from, say, the manipulation of thematic and generic expectations from the opening lines of F. T. Prince's poem 'An Epistle to a Patron' (see below, pp. 200ff). But in between these micro- and macro-components of a text – from the accumulations of alternative potentials for meaning in a single speech from *Hamlet* (see above, pp. 106–10) to repetitions with variations of the same narrative material at different stages in the progress of a complete narrative in *Tristram Shandy* (see below, pp. 192–200) – the references of the textual material to an intention rendered intelligible by the application of determinate concepts is obviously inappropriate. The poem needs the conceptual raw materials to create conditions within which the free play of what Kant calls the understanding and the imagination can take place. A degree of complication appears to be necessary for this non-cognitive free play to get under way.

For Kant poetry was first among the arts just because of the combination in it of conceptual but non-sensational raw materials, and the containment of these materials within the non-purposive framework common to all the arts. On the whole, 'The Analytic of the Beautiful' suggests that in art Kant entertained a preference for dependent over free beauty. Indeed, there are passages in which he seems to think all artistic, as distinct from natural, beauty is dependent beauty.[64] But the history of Idealist aesthetics and Romantic and Symbolist poetry throughout Europe after Kant's death entirely reversed this order of priorities. The important Kantian requirement that the purposiveness of a work of art should be indeterminate, owing nothing to the power of the will and being separate from the faculty of desire, was to be preserved. But the conceptual aspect of the verbal raw materials was to be as far as possible replaced by an emphasis on their merely sensational characteristics, most famously expressed in Baudelaire's theory of correspondences and Mallarmé's prose poems. Also, the distinction between the subjective finality of a natural object and the objective finality of a work of art became progressively hazier as the functions of words as bearers of meaning were to an increasing extent ignored.

It becomes clear, then, why, next to a rose or some other natural object, music rather than poetry should have come to take first place among the arts. That is to say, among the arts that display the essential characteristics of *free* beauty – purposiveness without purpose both in the overall final form of the work and in the particular, even the most basic, elements of structure that it comprises. For unless it is a tone poem or an operatic overture, music, like a rose, represents nothing but itself. Our subjective feeling about the way the movements of a symphony 'fit in' with one another is not in this respect altogether different from the way the petals of a flower are felt to do much the same thing. And it is true that, since the emergence of the Romantic movement at the end of the eighteenth century, music has occupied a special position in the hierarchy of the arts on account of its non-representational character. Kant's initially neutral distinction between free and dependent beauty has been converted into a value-laden distinction between different categories of beauty and a parallel distinction between different degrees of purity in our responses.

Kant made no such value judgements as are implied in the post-Romantic preference for *la poésie pure*, suprematism, or *musique concrète*.[65] He merely articulated a distinction between natural objects or works of art which evoke feelings of beauty unattached to any anterior or ulterior purpose that would imply a concept, and other objects (e.g. tools) or works of art which evoke feelings of beauty that are attached to such a purpose and that do, therefore, require a concept. This was to be a distinction developed in the expressive theories of Croce and Collingwood at the beginning of the present century. But as a matter of fact we happen to know that Kant's aesthetic taste inclined towards the dependent beauties of Milton, Haller and Pope. How could they have done otherwise, since his preference for literature over music and painting implies a preference for the impure? The point has already been made that literature, being constructed out of words, must refer to concepts. It cannot, like music or painting, be entirely abstract. It is bound to be less like a rose than even a tone poem by Schumann or Strauss is like a rose. Together with concepts, words must imply purposes. Sounds and shapes are continually buzzing in our ears and appearing before our eyes, without provoking us to wonder about the purpose and intention of their being there. But words always express purposes. They are man-made, unlike the colour blue or the sound of F-sharp. And men are purposive creatures. Words are always for something and about something; blue is just blue. Blue is a word that stands for a colour we can respond to without giving it a name. F-sharp is the way we represent in the form of a note a sound we can hear without thinking of it as a note, let alone as a note defined by the use of the words 'F-sharp'. No doubt when a composer assembles his F-sharps in an arrangement that includes many other notes representing many other sounds, he is performing a series of intentional acts that conceal a purpose. But the purpose is much more easily concealed where materials that are in some sense purposively handled are not themselves members of a purpose-laden category of object.

Applying Kant's aesthetic teleology to the study of literature, we are bound to encounter difficulties. How can a text manifest the free beauty of purposeless purposiveness at the same time as it is bound to engage its readers in the application of concepts – because the concept-bearing material out of which it is formed, words, is too undisguisedly and brazenly apparent? Commentators usually answer this question by referring to Kant's aesthetic ideas. Their presence in works of literature accounts for the conceptual content of the words and the meanings that combinations of words contrive to spread over the surface of the whole poem, play, *etc*. This is acceptable until the point is reached where the value of a poem begins to be accounted for by the capacity of those meanings to enter into a relation with practical reason, thus conferring on them a sort of supersensible reality. (This was the approach to Kant that made Coleridge so selective in his borrowings from *CJ*.) For this reason, I have excluded aesthetic ideas as far as possible from the argument. It hasn't been possible to do so completely because, in a much more humdrum way than Kant has been (correctly, but selectively) interpreted as suggesting, they are bound to enter into our reading: it is both impossible and undesirable to deny that words combine in poetry to signify in ways they don't signify in other kinds of discourse as well as in ways that they do. Scruton is right to emphasize the way the poetry articulating the vengeful feelings of Satan in *Paradise Lost* conveys to us 'not this or that, as one might say, "contingent" emotion, but the very essence of revenge'. He is right, too, in being circumspect about the extent to which the aesthetic *experience* of listening to those words is to be detached from the aesthetic *ideas* they 'seem' to incorporate. Coleridge followed Kant in arguing that 'aesthetic ideas...are ideas of reason which transcend the limits of possible experience, while *trying* to represent, in "sensible" form, the inexpressible character of the world beyond' (Scruton's words, my italics). But he refused to accept the force of what the word 'trying' tells us in that sentence. Again, Scruton rightly emphasizes that while it is true that 'aesthetic *experience*, which involves a perpetual striving to pass beyond the limits of our point of view, seems to "embody" what cannot be thought', the word 'seems' is very important, because it insists on the fact that at this stage in one's reading of the poem, concepts have been left behind.

Two things need to be done. Kant's aesthetic ideas need to be demystified in something like the fashion he suggests and that Scruton picks out in the emphasis he gives to words like 'seem' and 'trying'. Then what is left of them needs to be reinforced with what we can discover in the content of the words that conforms to the idea of purposive wholes being wrought out of the various kinds of intentional activity I have been trying to describe.

McCloskey gives an indication as to how this might be done – indeed how Kant himself seems to have hinted we might try to do it – in the contrast she makes between 'proximate' and 'ulterior' ends. Just as musical notes and strokes of paint are the raw materials of some of the other arts, so the sounds *and senses* of words are the raw materials of literature. Combinations of sounds, in verbal patterns, are bound to bring with them combinations of sense, often sense that has a conceptual content. According to McCloskey's interpretation of Kant, these would now become the vehicles of a dependent beauty which makes aesthetic ideas universally communicable. The ulterior end is the communication of aesthetic ideas, realized in the achievement of the proximate end of assembling the words into conceptually charged combinations of sound and sense. We have to replace those aesthetic ideas, as the ulterior end of writing, with a 'fit' between the accumulation of intentions in doing and the indeterminate sense of purposiveness attributed to the world that is reflected in the work of literature that such an accumulation brings into being.

The appeal of Holloway's 'overall perspective', articulated in terms of intentional and purposive activity, is always in danger of being obscured by the very content and detail that has to be used to project it over the surface of the poem. It needs the reinforcement of what are in the long term (sometimes the not so long term) ephemeral matters of interest, extending from the range of meaning of particular words, to the widest possible cultural assumptions. This is why Wittgenstein thought authors became dated, 'even though they once amounted to something'. Their writings, he said, 'when reinforced by their contemporary setting, speak strongly to men, whereas without this reinforcement their works die, as if bereft of the illumination that gave them their colour.'[66] The fact that it is one of the principal functions of literary scholarship to recuperate these ephemeral interests and assumptions, and that many people – perhaps most – read the classic authors in order to recreate for themselves the times in which those authors lived and the moral, social and political issues that preoccupied them, should not blind us to the greater importance of the fact that these are all secondary, because dependent, considerations for the reflective critic. He knows, as other readers do, that novels and plays and even poems are chock full of these perfectly legitimate sources of interest. More than that, he knows that in many cases the author wrote those texts in order to articulate a very precise and determinate private or public purpose. Hence, everyone knows that Dickens approved of the Victorian reformers and Tolstoy the Russian peasantry. But it is one thing to establish these banal facts about *Oliver Twist* and *War and Peace*, quite another to establish the extent to which their presence in these books accounts for the aesthetic pleasure we derive from reading them.

Their status in the present argument is in one sense a very elevated one, in another not elevated at all. Poems can't do without the sort of purpose and content attributed to Dickens and Tolstoy because words have meanings and combinations of words express concepts and represent events. But this is to take a very negative view. Clearly there is a less grudging sense in which it is important that *Oliver Twist* is about the Poor Law and *War and Peace* is about Napoleon's invasion of Russia. On the one hand these issues have an intrinsic human interest that the exercise of a little historical imagination will make immediately apparent, and on the other they provide opportunities for much reflection on and discussion of matters of universal concern. T. S. Eliot[67] had Wordsworth in mind when he claimed that it was impossible to leave 'wholly out of account all the things for which he cared deeply, and on behalf of which he turned his poetry to account', and no one was more sceptical than Eliot of the importance to be attributed to the author's 'conscious purpose' ('The Music of Poetry') in the writing and reading of poetry. In fact every reader knows this is by no means the whole truth about the imaginative value of the poem's content. The difficulty has always lain in showing how the content combines with the form of the poem to produce its overall effect on the reader. I am suggesting that this kind of content is valuable in so far as it provides the conceptual and representational raw material out of which the creative possibilities of complex intentional activity are to be derived. At the same time I am suggesting that such raw material has an intrinsic interest. It is a condition of the possibility of the use of such material that it should possess the sort of ideological and human interest readers will be prepared to attend to and writers will find sufficiently ductile to shape into the elaborate intentional structures of literary art. In what follows, I want to hold on to the commonsense view that we know much good literature was written with an immediate representational, purposeful, sometimes even propagandist end in mind. At the same time I want to subordinate this knowledge to considerations of a less brutally 'dependent' kind. It should be possible to contain what at first sight might look like overriding purposes of the sort represented by Dickens and Tolstoy within something resembling more closely the purposiveness without purpose Kant saw as the essential fact about beauty in both its free and dependent state.

INTENTIONALIZING KANT

> Forget not yet the tried intent
> Of such a truth as I have meant
> Sir Thomas Wyatt.

To do this we must return to the equation made on pp. 165–6 above between intentions of doing and purpose, and intentions in doing and purposiveness. What does this mean in the context of Kant's distinction between free and dependent beauty subject to pure and impure judgements of taste? As I have indicated: 'A character can have a purpose without having a specific end in view. But an intention is deliberately...end-directed' (p. 80). This applies to both a writer and his characters. Now I have made the further distinction between purpose and purposiveness, and intentions of doing and intentions in doing, which, I hope, help to clarify how this comes to be so.

Intentions of doing, or intentions to do, correspond very closely with intentional motives. A character in a book, or a writer of a book, as a consequence of psychological motives with which we need not concern ourselves proposes to himself a course of action that may be very definite – in the sense that it is predicted that a particular effect will be achieved by it – or it may be much more amorphous and open-ended. In the case of the character Iago, the course of action he proposes to himself is to get Cassio's place (definite) and to plume up his will (amorphous). In the case of the author Shakespeare, he launches the ghost into the story of *Hamlet* in order to incite Hamlet to revenge (definite) and to induce in Hamlet a state of mind in which he might do any number of things or he might not do them (amorphous). An intention of doing, i.e. an intention to do something definite, will remain neither more nor less than that until obstacles of one kind or another are placed in its path, forcing it to redirect itself in accordance with a different aim that is not interrupted by any such obstacles. At each point (only one point if no obstacles appear) an intention of doing looks forward to a specific goal or end. To use the word 'end' in this context is the same as to use the word 'purpose'.

Purpose is related to an intention of doing as a target waiting to receive an arrow is related to the bowman's aim as he releases the arrow from his bow.[68] They are the same thing viewed from different sides of a motivated action. When a person says 'my purpose is to do x' and when he says 'my intention is to do x', he means exactly the same thing. The difference is not one of meaning but of the angle from which the meaning is observed. When he speaks the first sentence he looks forward to what will be accomplished and sees whatever that is in relation to the 'curve' of the intentional action he expects to accomplish it. When he speaks the second sentence he looks forward along the curve of the intentional action towards what he seeks to accomplish and sees the curve stop at the point of accomplishment. In its extreme form, this tight 'fit' between an intentional act and its foreseen realization is dramatized in the behaviour of Sophie Stanhope in Golding's *Darkness Visible*. Throwing stones at dabchicks on a stream:

> ... she saw the curve which the stone would follow, saw the point to which the particular last chick would advance, while the stone would be in its arc... When she thought back later it did seem that as soon as the future was comprehended, it was inescapable... as if... foredoomed from the beginning, ... as if the whole of everything had worked down to this point – to follow that curve in the air, the chick swimming busily forward to that point...: then the complete satisfaction of the event.

T. S. Eliot writes in the same terms about the way the reader traces the flight of the poem to its target: 'endeavour[ing] to grasp what the poetry is aiming to be; one might say... endeavouring to grasp its entelechy.'[69] Writers, too, have intentions and purposes of this kind. They also are in the business of connecting aims and entelechies. Author X, having the intention of doing y, will have y as his purpose. But as I have suggested, writing a book, like plotting against Cassio or Othello, is a complex action – not at all like shovelling snow with the intention of paying the builder. For the writer, it is more likely to be a bank holiday, the tyres won't grip on the ice and the dark is coming down over unfamiliar roads long before 4.30 on a Friday afternoon.

Literary intentions tend towards the amorphous. They are more likely to stray and to change in the course of time than most other intentions do. As they become more amorphous, and are seen to become so by the reader, the reader becomes more preoccupied with the writer's intentions in doing what he does than in his intention of doing it. Intentions in doing tend to have a much less clearly defined objective than intentions of doing, because they have moved further away from their moorings in intentional motive. This makes a lot of difference to the connection between intention and purpose. If the intentional act has become less deliberately and less accurately aimed, the target or goal of this activity will become correspondingly less clearly visible, less real in its character as the purpose towards which the intention in doing tends to move.

So what kind of purpose does an intention in doing serve? To what end does it aspire? The question is difficult to answer, because so much depends on the point of vantage from which the servant or aspirant is viewed by the person who asks it. In asking about intentions of doing, the questioner stands beside the intending subject at the moment he defines his intention, or proposes to himself a course of action. That is why Searle uses the phrase 'prior intention' to describe it. For intentions of doing are always forward-looking. An effect is predicted which the activity that follows the prediction seeks to bring into being. On the other hand the point of vantage from which one asks what is the intention in doing is not easily and unambiguously located at either end of the series of intentional actions. It will always be *in medias res*, in the midst of things. And because that is so, its position will always be less firmly fixed in one place than that of its twin will be.

Another characteristic of these kinds of intention is that they often exist in close association with each other. An intention in doing will often turn out to be an intention of doing *in a context*. The answer to a question of the 'what is your intention in doing x?' variety will usually take the form of 'my intention in doing x is to achieve y', or 'that of bringing about y' (which is merely a disguised 'to...' expression). One could define 'intention in' as the provision of intentional contexts within which voluntary and purposive acts are performed. But often such provision is not made and the 'intention to' stands out boldly as the occasion for and explanation of whatever acts are to follow. At a later stage in the description of the purposive activity such an intention of doing seeks to govern, the context may or may not be provided – in the form of the 'viewer's' (probably the author's) interpretation of the agent's intention in doing what he does. Where this happens, in the process of recording what is happening as present intentional activity, a hiatus will inevitably be produced between the original intention of doing and the purpose that was originally foreseen at the time the intention was formulated. This will happen because less authoritative and more short-term purposes will have interrupted the free realization of the original intention of doing.

In other circumstances intentions in doing, far from providing an intentional context for voluntary and purposive acts, pull loose from intentions of doing altogether. Here, the answer to a question of the 'what is your intention in doing *x*?' variety will be something like 'I did it, that's all', and that will be to describe an intentional action, though one that includes no reference to a governing intention. Searle says of such a case that 'the intention was in the action but there was no prior intention', and again:

> We say of a prior intention that the agent acts on his intention, or that he carries out his intention, or that he tries to carry it out; but in general we can't say such things of intentions in action, because the intention in action just is the Intentional content of the action; the action and the intention are inseparable. (p. 84)

Relating these distinctions between intentions of and intentions in doing to the distinction I have already made between specific purposes and a general sense of purposiveness, I want to claim that where the intention in doing provides a context for intentions of doing, it might deflect that intention from its aim of realizing a specific purpose; and where an intention in doing is presented without reference to an intention of doing at all, it will appear either as a random act or as a link in a chain of purposive activity where, and only where, it can be seen to be related to other acts with the same un-purpose-laden but purposive characteristics.

In the first acts of Shakespeare's tragedies considered above, events occurred as a result of one or more of the characters' intentional activity. But I was at pains to distinguish between (a) intentions – like Macbeth's intention to kill Duncan, which, at an early stage in the play, is definite, active and conclusive; and (b) purposes – like Iago's in 'working on' Othello, which fails to assume a definite or conclusive shape until the events of V. ii have taken place. When I turned from the characters to Shakespeare, I made a similar distinction. Shakespeare used the ghost in Hamlet to convey to the hero his duty to avenge his father's murder by killing his uncle. That was what accounted for his intention to introduce a ghost in the early scenes. Like Macbeth's, it was definite, unambiguous and ought to have been conclusive. It was not conclusive because Hamlet's intention *of* using the ghost to achieve this simple aim was complicated by his intention *in* using the ghost as he did. This had the effect of placing him in a free state where the uncertainties and doubts festering in his mind before he saw the ghost were compelled to take an active and dramatic form. In its turn this had the effect of supplying the play with more opportunities for varied and uncertain development than it could ever have acquired had it been governed merely by Shakespeare's intention to have the ghost drive Hamlet to his revenge.

Plays like *Macbeth*, *Othello* and *Hamlet* work as well as they do because of the way the author's purposes and purposiveness are reflected in the purpose-laden behaviour of his characters. There is a dynamic interplay between Shakespeare's and his characters' intentions to do certain things, and Shakespeare's and his characters' intentions in doing them. Purpose and purposiveness grow out of each other in ways that are unpredictable – to the characters and, because to the characters, to Shakespeare also. The plays develop into purposive wholes which, like all art that expresses Kant's free and dependent beauty, have no purpose. But they could not have been purposive in this way if purposes had not been visible within them. Those purposes are the manifest appearance of Shakespeare's intentions of doing what he chose to do, built into the behaviour of characters who act as his surrogates and enablers in converting limited purposeful activity into illimitable and inexplicable purposive form.

What happens, though, when the form is not dramatic, and the control exercised by the writer is more transparent? Suppose, instead of a tragic play, one were to choose a narrative poem or a novel. Wouldn't this make a lot of difference? Wouldn't the author's hand be exposed in the construction of the episodes and the divisions of the chapters? If it were so exposed, wouldn't it be seen to be organizing the fictitious events for a purpose, with the intention of achieving a predictable effect by recourse to methods that were intermittently or even permanently before the reader's eye? In other words, wouldn't the admired Shakespearean freedom have to give way before restrictions imposed by a governing purpose that might be to a greater or lesser extent explicit, but that would in either event insert itself between the free development of the narrative and the reader's expectant response?

Generic conventions always make a difference. My reading of *Othello* would have placed the emphasis differently if it had been preoccupied with the derivation of Iago's character from the Vice of the Morality plays and the anonymous Ensign of Cinthio's novella. One would have thought, then, that the play of intentional force would manifest itself differently in what are not merely different genres, but different forms of imaginative writing themselves acting as host to a wide variety of generic *topoi*. Bearing this in mind, and comparing Shakespeare's plays with narrative poems, novels and verse monologues from the seventeenth to the twentieth centuries, a reader might expect that an author of non-dramatic forms of narrative must be at a severe disadvantage. His access to the characters – whose intentional activity generated so much of the sense of free, though purposive, movement in the plays – is blocked by the presence of all those descriptive, ratiocinative, psychologically interpretative or otherwise explanatory passages that account for the larger part of most narratives. And narratives that are largely or entirely constructed out of dialogue, like the novels of Ronald Firbank or Ivy Compton-Burnett (I set aside the later Henry Green as an exceptional case), tend to draw attention to the rigorous control of the author's ulterior intentions at least as markedly as those that observe the traditional conventions of balance between dialogue and description and commentary.

Nevertheless, we have learned enough in recent years about the essentially insincere nature of fictional representation to look with some scepticism on an approach to any genre of literature that makes naive assumptions about the author's 'presence' in his work. The danger of personal intervention is obvious and infrequently resisted. But in much good fiction an artificial surface is provided which can work to the advantage of the writer where this matter of converting intentions of doing into intentions in doing – that is to say of converting purposes into purposiveness – is concerned.

I want to elaborate on the ways writers achieve the Kantian aim of producing the illusion of purposiveness without purpose by looking at two different specimens of narrative in non-dramatic form. First I want to look at certain features of *Paradise Lost*, with a view to determining how Milton – though he is telling a story which, as he announces at the beginning of the poem, is intended to serve a limited (though grand) and specific purpose – manages to do so in a way that removes the intentional 'sting' from the procedures he adopts to achieve his aim. At first *Paradise Lost* looks like a closed system, a narrative poem whose *raison d'être* is to realize an intention to do something very specific. I want to show how that system opens up and opens out, creating possibilities for that free movement of the imagination which, in Shakespeare, I attributed to the transformation of end-directed intentions into intentions in action that discover ends they do not appear to seek.

Then I shall turn to what at first may seem to be the very opposite kind of narrative: the endlessly digressive and meandering story of *Tristram Shandy*. Here what I have to resolve is the reverse of the problem encountered in *Paradise Lost*. For in *Tristram Shandy* we encounter such a weak grasp of purpose, manifesting itself in such an aimless drift of ill-defined intentions and apparently pointless gestures, that it is difficult to make out any sort of shape at all to the author's intention in writing the novel. It is not that Sterne's characters have no firm intentions or clearly defined sense of purpose. They do – sometimes to the point of mania. But in ways that will be apparent to any reader, they fail to let this rub off on their author, and consequently seem to achieve very little in spite of their frantic efforts in this direction.

In *Paradise Lost* the higher purposiveness both mimes and undermines Milton's didactic activity, energetically and vigorously displayed in the scheme of the epic poem. In *Tristram Shandy* it mimes and reconstructs Sterne's baffled inactivity, thwartedly and regressively displayed in the 'scheme-lessness' of Tristram's failure to write his autobiography. But in both books – poem and novel – most readers, I think, have felt the presence of that purposeless purposiveness that Kant said was a salient characteristic of free and dependent beauty in all great art.

4
Milton, Sterne, Prince

> We may not think the justness of each act
> Such and no other than event doth form it
> *Troilus and Cressida.*

THE INTENTIONAL GRAMMAR OF *PARADISE LOST*

> Language always includes agency, and agency and intention are frequently impossible to distinguish in language.
>
> Gillian Beer, *Darwin's Plots.*

Discussion of *Paradise Lost* has been bedevilled by confusion in the critics' mind between psychological and intentional motive. What Milton wanted to do, and what the terms of his setting out to do it permitted him to do, have been interpreted as two different – and perhaps incompatible – things. Consequently, reading the poem has become a difficult and complex activity in a sense Milton surely never intended it to be.

Unprejudiced readers are bound to attach a great deal of importance to Milton's announcement to his Muse, at the end of the first paragraph of *Paradise Lost*, that his 'great argument' is to 'assert eternal providence, And justify the ways of God to men'. This is the way the poet has chosen to announce his prior intention, his intention of doing something in writing the poem. Since this statement of intention appears just 25 lines from the beginning of the first book of a poem that will eventually exceed ten thousand lines, I can't feel that Milton's having such a prior intention is seriously compromised by a narrative context that has not had time to influence the single-minded deliberation with which the poet makes his point, or the simple-minded and open responsiveness with which his audience interprets it. But I want to stress that what Milton announces in this invocation *is* an intention. At no point does he raise the issue of psychological motive, revealing to the Muse or to ourselves that he intends to do *x* because of *y*. Even in his more personal-sounding invocation at the beginning of Book III, he doesn't refer to his blindness as a motive for undertaking his task, or for performing it in the way he is doing. On the contrary, his physical blindness is brought into the poem to add force to his plea to the Holy Spirit to 'Shine inward... that I may see and tell/Of things invisible to mortal sight.' (lines 52, 54–5) – in other words, to add force to the plea that his intention shall be realized in the poem ('see and tell').

His intention is to describe an action. No one to my knowledge has argued that asserting eternal providence and justifying the ways of God to men are two different and unrelated things. The 'and' here is performing the opposite function to Iago's 'and' in his first soliloquy in *Othello* ('I hate the Moor, and...'). It means that eternal providence will be asserted through the description of God's actions in relation to men, which themselves will be justified in terms of the description. This is bound to be a description of acts one would expect to be in the strongest sense of the word 'intentional'. We have looked at different kinds of intentional activity performed by Macbeth, Iago and the ghost in *Hamlet*. How much more intentional, though, must the acts of our first parents have been. And how much more so must have been those of the supernatural powers whose fortunes became entangled with theirs. The consequences of the intentional acts these characters performed stretched (literally) infinitely further beyond themselves than we can imagine any intentional act of a merely historical or entirely fictional figure like Macbeth or Othello having done.

There is another difference, to be taken notice of along with the difference in the strength of meaning of 'intention' in descriptions of intentional acts in *Paradise Lost*. In all the characters from Shakespeare, intentions were closely associated with motives. Shakespeare's own motives might not be at all to the point. In any case we have no idea what they were. But Hamlet's were very much to the point, and the connection, or lack of it, between his motives and his intentions lies close to the heart of the play. In *Paradise Lost* we rightly take an interest in Adam's and Eve's motives, especially from Book IX onwards, where Milton announces that he has changed his notes to tragic and is viewing his human protagonists as fallen beings. But as we move from mankind to Satan and the fallen angels, thence to Gabriel, Michael and the unfallen angels, to Christ and finally to God the Father, we move further and further away from a psychological universe where considerations of motive are appropriate.

At the same time we move away from characters and circumstances that almost all readers of the poem since Blake have found interesting and successfully realized, to characters and circumstances that the same readers have found uninteresting and poetically weak. This must have something to do with the fact that we associate so closely in our minds the successful representation of characters in action and motivated activity. So long as intention is more or less identified with motive (as in Anscombe's 'gain'/desire of gain' equation) such association is bound to persist, because it is difficult for us to conceive of a state of mind in which intentional motives are not governed (even if at a subconscious level) by prior psychological motives. For example: I am in love with my mother (subconscious psychological motive), therefore desire to kill my father (prior intention or intention of doing), and therefore kill my father (intentional action). But in his representation of the angels and Christ, and still more of God the Father, Milton is trying to give poetic expression to intentional motive cut off from psychological motive, which means intentional activity not grounded in psychology or motive at all. The assumption that this is impossible or, if not impossible, that Milton was ill advised to try and do it in an epic poem accounts in large part, I think, for the critics' hostility to Milton's representation of God.

Problems attending the representation of the Christian supernatural have been couched in terms of motive, intention and purpose long before the philosophy of action came onto the scene. For centuries they have exercised the minds of reflective men who were not poets and who lived long before the Romantic movement changed the terms in which most other literary-theological issues were discussed. For example, theologians have wrestled with the problem of how motives and intentions expressed in biblical descriptions of God's relations with man might be understood in rational-discursive terms, and among these theologians none did so more strenuously than the Calvinist divines who were Milton's contemporaries and immediate predecessors. It would be possible to examine Milton's text in something like these seventeenth-century terms, referring to the substance of contemporary theological debate, and placing at the centre of the argument matters like voluntary and involuntary acts of faith or modes of reprobate behaviour in relation to the exercise of free or variably constrained moral choice. Such an examination would include references to controversialists like William Ames and Thomas Hooker, whose writing is much preoccupied with these matters.[1] Their treatises and sermons established the terms and reinforced the mind-sets of the participants at the Westminster Assembly of 1643–9, and subsequently the articulation of Anglican soteriology in the Confession of Faith that Assembly produced.

Milton was writing *Paradise Lost* in the wake of these controversies. Clearly he shared the interest they had awakened in representing the dynamics of God's intentional plan and man's response to his role within it as both humanly intelligible and logically and morally defensible – from both a human and a superhuman point of view. But his ambition to render these 'events' first in dramatic, then in epic narrative brought with it both a challenge and a problem. The difficulties of making God's plan intelligible in the terms appropriate to the composition of an epic poem were very different from those involved in the discursive and rhetorical modes adopted by contemporary theologians or in his own *De Doctrina Christiana*. Milton encountered similar difficulties to those encountered by contemporary theologians, but he had to find different ways of resolving them in an epic poem. Readers have explained the peculiar mixture of delight and dissatisfaction they feel when reading *Paradise Lost* in a variety of ways. Tillyard writes about conscious and unconscious meanings, Stanley Fish about realizable and unrealizable linguistic aims.[2] But the issue is at bottom the same, and it is related to the ways Milton went about resolving these difficulties. How else explain the correspondence between the ascending curve of critical dissatisfaction and the ascending curve of the theological status of the characters represented in the poem? Critical dissatisfaction grows more intense with each ratcheting upwards in theological status, and each ratcheting upwards in theological status is accompanied by a noticeable diminution of the psychological motivational component in the intentionally motivated activity.

This is what might at first appear to be the unfortunate consequence of implicating Milton's failure with God's success. For Milton aspires to see the world, as far as it is humanly possible, from the point of view of God: he intends to soar 'with no middle flight', and asks the Holy Spirit to 'raise and support' him at a 'highth' where 'heaven hides nothing from thy view'. At a linguistic level, then, Milton favours a mode of expression which, when addressing itself to descriptions of the highest Being, takes on the colouring of that

Being's existential character. Much of the point of the poem, as Fish would say, lies in what is ultimately Milton's inability to do this. That is what the Fall, linguistically considered, is all about. Human language has deteriorated along with human understanding and human morality. Nevertheless, poems are fantasies, not tracts for the times. So in those parts of his epic fantasy where God himself is represented as speaking, Milton does the best he can to mime poetically how God reveals himself existentially. (Essentially, God wouldn't be describable at all.)[3] To do this he has to find a language that separates psychology from intention, leaving God active at the level of intentional motive, but quiescent – or, as it were, positively non-Being – at the level both of psychological motive and intentional act. Somehow or other, intentional motive has to be made to isolate itself, linguistically, from everything that is commonly and necessarily supposed to precede it as cause and succeed it as effect.

When God first appears in *Paradise Lost*, at III. 56ff., he does something which, for him, is very unusual. In fact I don't think he does it again in such a straightforward and uncomplicated way anywhere else in the poem. He performs an intentional act:

> Now th'almighty Father from above,
> From the pure empyrean where he sits
> High throned above all highth, bent down his eye,
> His own works and their works at once to view:

Allowance having been made for the way Milton's syntax has the effect of separating subjects from their actions here as elsewhere in the poem, what Milton is saying is that the almighty Father *bent down to view* what he had made. Referring back to my discussion of Hamlet's 'To be, or not to be' speech (see above pp. 106–10) the reader will at once appreciate that this is one of those purposive infinitives where the 'to' prefix actually does mean 'to' or 'in order to'. It expresses an act of will, an intention to be realized in action. This is an unusual if not unique application of this grammatical form in all of those parts of *Paradise Lost* which either describe God from a disinterested, or angelic, point of view, or reproduce his speech in the speech of another character. It is easy to misrepresent what the poem actually says in even the most scrupulous paraphrase of these parts of *Paradise Lost*. For example one recent commentator paraphrases VII. 548–57 as follows: 'After that creative act, God ascends "unwearied" to Heaven to behold the "new created World" and to see "how good" and "how fair" it is, "Answering his great idea".'[4] The placing of 'Thence' between 'up returned' and 'to behold' removes much if not all of the verb's intentional force, insisting on the sense 'where he beheld', i.e. a simple adverbial clause of place in the infinitive form.

The only other place where the purposive infinitive appears in such a pure form is in the 'Argument' of Book V, where Milton writes that 'God to render Man inexcusable sends Raphael to admonish him of his obedience'. I don't think I am alone in finding this description of God's intention uncharacteristically deplorable. The blunt and savage expression of intention here shocks in a way God's considerably more elaborate, even tentative statement at III. 80–134 does not. I confess myself utterly baffled about Milton's intentions in letting these words slip into a passage of prose so intimately associated with the poem.

(The infinitive in the 'Argument' of Book V is a formal indication of psychological motive that scarcely ever appears elsewhere in God's speech in *Paradise Lost*. There are one or two instances of other characters, especially the devils, imputing psychological motive to the Almighty: e.g. Belial at II. 156–60, and Satan at V. 781 and 424–5, VI. 625, and IX. 143–51, 176 and 703–4. Also, Satan endows God with intention at IV. 898. On just two occasions the Son and the good angels attribute infinitive purpose to God, but on neither of them with any clear attribution of motive: see VI. 724–5, and VIII. 239–40. With the exception of the passage quoted above, God himself speaks infinitively at III. 279, V. 228 and 232, VI. 40 and 675, VII. 'Argument', 147–8 and 554, VIII. 343–4 and 348, X. 55, 62 and 629–30, and XI. 95. God actually uses the word 'intend' at VII. 446 and at X. 58. In all of these cases except the ones from Book VII [where the divine infinitive does express intention, though not motive], there is either some doubt about the extent to which God is credited with performing an intentional act or some peculiarity about the grammatical form in which he articulates it. Some of these passages are discussed below. I offer the complete list here so that the reader can refer to examples from elsewhere in Milton's text.)

God's purposive viewing at III. 56ff. is as little intentional as an intentional act can be. It has no effect, can have no effect, on anyone but God himself. No doubt he is satisfied with what he sees. The angels 'from his sight received / Beatitude past utterance', but that is more on account of their 'sighting' of him than his sighting them. It is the sight of him that strikes them dumb. The fact that he is viewing them when sighted is incidental to their feelings of beatitude. How different is Satan, viewed by God in the same verse paragraph, totally unaware of 'Him God beholding from his prospect high', and with firm purpose 'ready now/To stoop with wearied wings, and willing feet,/On the bare outside of this world'. There is an immense contrast between Satan's sense of purpose expressed in a specific activity as a result of which a particular desired effect is expected to materialize and God's inactive presence even where he is described as doing something deliberate: 'viewing' his creation. The stillness of God, contrasted with the activity of all the other characters, has frequently been commented on: 'Metaphysically He would seem to be properly outside the universe of action, though He is also the biblical God, revealing Himself in His chosen terms, however symbolically, to angelic and human understanding.'[5] I am responding to these lines in their linguistic character. It may be that God is intending to do all sorts of things and that he is on the brink of realizing them by performing any number of actions. But at the level of language, even when we do have the purposive infinitives to take into account, he is much less intentionally active than Satan. Elsewhere we shall find that this is even more strikingly the case, because these purposive infinitives will usually attach themselves to anyone *but* God, and especially to Satan – and Eve, the most intentionally active of the two human characters in the poem.

Later in his speech God makes liberal use of the infinitive of purpose, but he never applies it to himself. Both Satan and Man are the subjects of intentional actions, though we know that responsibility for them must ultimately be traced back to God. After all, he is the author of all things. Indeed he comes close to insisting on this when he says of Man that he (God) 'made him just and right,/ Sufficient to have stood, though free to fall' (Book III, 98–9). These memorable lines sound like a statement of intention on God's part. But in fact, grammatically, they are not. The insertion of the two adjectives after 'I made him' ensures that the meaning we are expecting to hear (something like 'I intended him to stand or fall: the choice was his' – but with the idea of *making him do* something) is transformed into something that sounds as if it is very much less dependent on what God does. God makes man. Then man is empowered to do one thing or its opposite, where the empowering words ('Sufficient to...' or 'free to...') are so far detached from God's agency as to divest God of any specifically intentional activity at a linguistic level. That is what I mean by saying that God's intentional motive is freed from any poetically comprehensible intentional activity. Satan, in the same speech, is full of such intentional activity: 'with purpose to assay/If him by force he can destroy.' God says it, but Satan does it. So the sense of purpose sticks to Satan and seems to have nothing to do with God.

The detachment of purposive activity from God and transference of it to another, more psychologically motivated subject appears even more extraordinary when that subject is not the evilly scheming Satan, or even the irresponsibly self-willed Eve (or Adam), but Christ, God's own Son, and, in the present context, not only the corporeal representative of his mercy but also the verbal representative of his purpose.

As God pronounces his decrees verbally through the Word, the Son is both animated and becomes the means by which things are created. The Son as anointed and created offspring of the Father is thereby synonymous with the Word or utterance, God's verbal offspring. Deific utterance is of its nature creative: the Word itself is not only begotten, that is consecrated and guaranteed, but is the source of existence for others.[6]

The habit of linguistic transference can be seen to best advantage later in Book III, at lines 397–410. Milton is describing (more briefly than in Book VI) the angels' admiration of Christ's leadership during the war in Heaven. But he soon returns to the main business, which is to contrast this martial activity in the past with the more passive role Christ expects to play in the future:

> Back from pursuit thy powers with loud acclaim
> Thee only extolled, Son of thy Father's might,
> To execute fierce vengeance on his foes,
> Not so on Man; him through their malice fallen,
> Father of mercy and grace, thou didst not doom
> So strictly, but much more to pity incline:
> No sooner did thy dear and only Son
> Perceive thee purposed not to doom frail man

> So strictly, but much more to pity inclined,
> He to appease thy wrath, and end the strife
> Of mercy and justice in thy face discerned,
> Regardless of the bliss wherein he sat
> Second to thee, offered himself to die
> For man's offence.

Earlier and later lines are not concerned with willing and deciding. They are almost entirely given over to descriptions of God the Father and of his Son as 'divine similitude' and therefore contain no purposive syntax at all. The reference to the war in Heaven begins four and a half lines before the lines quoted, but since it applies to action alone, separated from notions of purpose or intention, there also no purposive syntax is required or expected. But the passage is framed by very bold examples of those purposive infinitives we found God was so sparing with. The Son, plainly seen here as the instrument of his Father's power ('Son of thy Father's might'), is extolled by the angels in order that he will avenge his Father ('To execute fierce vengeance on his foes'). The lines on Man that follow make it clear that the angels are looking forward to future vengeance, not back to vengeance already taken through the defeat of the rebel angels. So 'extolled...to execute' can't mean 'praised...for having already executed'; it must mean 'incited to execute in the future'. At the end of the passage three purposive infinitives, all related to the Son, follow one another in rapid succession. Christ, 'offered himself to die' 'to appease thy wrath' and (to) 'end the strife/ Of mercy and justice'. In all of these instances, Christ forms precise intentions: in the opening lines, at the prompting of the victorious angels; in the later ones, prompted only by his own Spirit after he has heard his Father's plan for mankind.

This is all very straightforward, and would not need to be described at such length were it not for what appears between the opening and closing lines of the speech. Here, from line 400 to line 405, the origins of purpose, and the responsibility for it, become much more intricately entangled. The rhythm and diction of the lines express this entanglement, even confusion, by reflecting and echoing each other even more than they usually do in Milton: 'Mercy and grace' (401) is half-reflected in 'mercy and justice' (407); 'thou didst not doom/So strictly' (401–2) in 'purposed not to doom frail man/So strictly' (404–5); and 'but much more to pity incline' (402) in 'but much more to pity inclined' (405).

Milton obviously decided to insert into the run of the blank verse a highly patterned sequence which has the effect of blurring divisions of responsibility between what is decided by the Father and what by the Son. It is done mainly by making the reader uncertain about the status of the word 'to', used three times here in four lines (402–5), but only once, on its second appearance, as an infinitive marker. That is, 'not to doom frail man' is the only phrase that actually contains 'to' plus a verbal infinitive expressing a sense of purpose. On the other two occasions, the repeated 'to pity incline/inclined' means that pity (of mankind) is something you incline to favour, not that you do something, i.e. 'incline' in order 'to pity' ('mankind'). The verbal suggestions of 'to pity' are reined in by the presence of the active verb, 'incline/(d)', which stops the line in each presentation. And since the 'thou' who governs this inclining is 'Father of mercy and grace', this means that in two cases out of three 'to'-clauses the language Milton uses has the effect of removing any specific purposive activity from God. I am reminded here of a critic's argument that Milton had to create uncertainty about which member of the Trinity is being referred to at key moments in the poem, e.g. VII. 207–9, where all three are named in the description of the procession from Heaven before the creation: 'One of the ways in which Milton portrayed the paradoxical unity and differentiation of the divine Persons was to confuse and blur the reader's distinction regarding who is acting...'[7]

What of the middle phrase, 'purposed not to doom'? There is no way of getting round the fact that God ('thee') is its subject. Should he not, then, be seen as having what is explicitly described as a 'purpose'? He should, were it not for the fact that the Son acts as the subject of the verb 'Perceive', which takes God ('thee') as its object. Then God, as an object of perception, has 'purposed not to doom' related to him as a verb in the passive voice. Or 'purposed not to doom frail man/So strictly' is a long adjectival phrase qualifying 'thee', as the object of 'Perceive'. This is very different from the 'purpose' used of Satan back at line 90: 'he wings his way.../ ...with purpose to assay...'. There the syntax displays Satan at the centre of an intentional action for which he is entirely responsible. Here the very different, more involved and hypnotically repetitive syntax displays God at the centre of a world of purposive activity which is enacted on his behalf by others. He appears to be untouched by the activity this verbal pressure sets in motion. He is authoritative and he is indispensable, but he does not perform any intentional acts of his own.[8]

The presence of 'not' in the phrase has the effect of draining off whatever purpose still exists in the verb. The same applies on a number of occasions when God's actions are described or his speech reported. A good example is at the opening of Book X, where we are told that 'God all-seeing' 'Hindered not Satan to attempt the mind/ Of man...' (6, 8–9). What Milton means here is that God deliberately didn't interfere with Satan's plans, because he wanted him to 'attempt the mind/Of man'. (It was part of the plan he explained to Christ and the angels at the beginning of Book III.) In the précis, Satan is the unwitting instrument of God's intentions. But in the poetry, he is the purposive plotter with whom God doesn't interfere. Even when God is represented as being purposive, he achieves purposiveness through the agency of a surrogate. If this is not Christ or one of the good angels, then it is one of those large abstractions – like 'Foreknowledge', 'Justice', 'dread Tribunal', 'high Decree/ Unchangeable' – that occupy so much of the space around him. At X. 13, the relevant phrase is the 'high Injunction' that Adam and Eve were warned they should remember: 'For still to know, and ought to have still remembered/The high Injunction not to taste that fruit.' Here 'not' strongly reinforces the meaning of the command. It doesn't have the effect it had in the earlier examples, of draining the sense of purpose away from the order. But since the purposive command issues from the agency of that 'high Injunction', God has once again retired from the arena of intentional action. For him, intention resides in the implications of epithets like 'All-seeing' and 'Omniscient'. It doesn't move directly through an intentional action to achieve an intentional object in which are discovered the conditions of satisfaction of that intention. This kind of activity is reserved for lesser beings whose perceptions and powers of will function in an environment where the distinction between prior intentions, intentional acts and intentional objects is more clearly marked.

That is why in *Paradise Lost* the lesser the being, the more intentionally active he will be and the more deliberately he will match his intentions to his purposes. Hence Christ is more active than God, Eve than Adam, Satan than the rest of the fallen angels. I mean syntactically active, in the purposive willing sense signalled by the 'in-order-to' infinitive, though as a matter of fact the characters tend to be more active in the more obvious dramatic sense of doing more as well.

Satan, the most damnably fallen character in the poem, is also the most busily active and the most syntactically purposive. He confers these properties upon the human actors with whom he comes into contact. At the moment before her fall, Eve avails herself more liberally than at any other time of the syntax of purpose: '. . . this fruit divine/. . . of virtue to make wise: what hinders then/To reach, and feed?' Adam, at the moment he resolves to join Eve in her transgression, strongly emphasizes the same purposive infinitive: 'Certain my resolution is to die.' His pathetic lines on the loss of Eve compound the alternative possibilities of purposes and result in the infinitive of 'How can I live without thee.../To live again in these wild woods forlorn?' (i.e. 'I must be without thee in order to live – in the only sense in which life, i.e. life-without-death, has any meaning for me – in these wild woods forlorn', or 'Living without thee would entail the result of living forlorn within these wild woods'). Most moving of all, at line 959, 'to lose thee were to lose myself' shows how Adam enters the fallen human condition by describing a foregone result in terms more appropriate to a forthcoming choice. The purposive ring of the two infinitives in a single line muffles the intellectual aberration of which he is guilty in claiming that to live with the fallen Eve is the only guarantee of life. In fact it is the pre-condition of certain death. As Adam said more truly (in his pre-lapsarian state) six lines earlier, by fixing his lot with Eve he is 'Certain to undergo like doom', which is death. That is why his equally 'certain' resolution was to die.

Satan's intense preoccupation with ends and purposes is reflected in his language throughout the poem. In almost every speech the purposive infinitive, conspicuously absent from the language of his great Antagonist, plays an important role. Take, for example, the following passage, from Book IV, lines 514–27, in which Satan's intentional inclination is vividly expressed in the syntactic arrangements of his speech:

> One fatal tree there stands of knowledge called,
> Forbidden them to taste: knowledge forbidden?
> Suspicious, reasonless. Why should their Lord
> Envy them that? Can it be sin to know,
> Can it be death? And do they only stand
> By ignorance, is that their happy state,
> The proof of their obedience and their faith?
> Of fair foundation laid whereon to build

> Their ruin! Hence I will excite their minds
> With more desire to know, and to reject
> Envious commands, invented with design
> To keep them low whom knowledge might exalt
> Equal with gods; aspiring to be such,
> They taste and die: what likelier can ensue?

There is an ascending scale of purposive content in the use of infinitives here. 'Forbidden them to taste' splits off the 'to' suffix from 'Forbidden' to which it properly belongs, by inserting the pronoun 'them' between it and 'taste'. This makes 'taste' sound a little bit more intentional than its sense demands. But the intentional or purposive colouring of the verb is nonetheless very slight. In 'Can it be sin to know...?', however, the 'to' prefix is packed tight against the infinitive as a result of its positioning after a noun, 'sin', and its absorption in the metrical foot at the end of the line. In the eighth line, the effect of the similar positioning of 'to build' is reinforced by the addition of 'whereon'. Though this adverb actually relates back to 'foundation', it can't be prevented from appearing to project forward to the infinitive, carrying with it a meaning closer to 'whereby'. The addition of the word 'desire' in 'With more desire to know and to reject' obviously loads the infinitive with a greater purposive energy, as does 'with design/To keep' on the next line and 'aspiring to be such' near the end. By these means Milton fixes in our minds the image of Satan as an unremittingly intentional agent, pitting his will against God in a flurry of purposive activity which has very definite ends in view. These are ends which, using Anscombe's terminology, we would define (and Satan would too) as how-intentions leading in fixed and foreknown succession to the final why-intention, or ultimate intention, of 'getting his own back on God'. He also allows the implications of Satan's rhetorical syntax to spill over onto Adam and Eve and, on one occasion (524–5), even attempts to do the same with God.

Satan is the ultimate intentional agent, the designer of intentions that entail one another with a remorseless logic. By deceiving the fallen angels he will appoint himself revenge hero; by appointing himself revenge hero he will traverse Chaos and alight on Earth; by alighting on Earth he will trespass in Paradise; by trespassing in Paradise he will enter the serpent; by entering the serpent he will tempt Eve; by tempting Eve he will bring about the ruin of mankind; and by ruining mankind he will get his own back on God. Maybe he doesn't plan all of these moves in advance: there is no mention in Hell of the serpent or of Eve. Beelzebub merely refers to 'some new race called Man' (at II. 348). The serpent disguise is an example of Satan's inspired improvisation: 'with inspection deep/ [Satan] Considered every creature, which of all/Most opportune might serve his wiles, and found/The serpent...' (IX. 83–6). But Satan does plan the main moves: to get his own back on God by ruining mankind by entering Paradise by getting the support of the fallen angels.

Although Satan is the source of all evil, the way he is evil is the opposite of Iago's way. As I argued above, Iago is a kind of artist in evil. His method is a wicked parody of Shakespeare's. But Satan is not an artist. He is more like a bureaucrat, or what people imagine bureaucrats are like. With this distinction in mind we might compare Satan's surfeit and God's lack of purposive activity. Satan's *non serviam* is itself articulated by recourse to the infinitive of purpose: who can rule us, he asks the rebel angels, who possess imperial titles 'which assert/Our being ordained to govern, not to serve?' (V. 801–2). We might look with renewed interest at Satan's hiding with a 'dark intent' in the mazie folds of the serpent (IX. 161–2), or, later in the same paragraph (176–7), at his imitation of Milton's phraseology in the 'Argument' of Book V. Here he attributes to God the intention 'us the more to spite' by raising Adam from the dust, and then secures for his own use the intentional act of spiting. The words with which he closes this speech are 'spite with spite is best repaid'.

In *Paradise Lost*, Milton gives his own distinctive twist to the idea that writers are like Gods, creating worlds that they either inhabit as moral legislators or from which they remove themselves to a fastidious distance – paring their fingernails like Flaubert and Joyce. But to the extent that Milton aspires to imitate his own God in the creation of his poems, he will not see these two alternatives as the opposite poles between which he is doomed eternally to range. For his own God lacks the psychological motivation of the first type and possesses in the fullest measure the intentional motivation discarded by the second. Since he is Omnipotent and Omniscient, there is no way God can be obliged to express this absolute intentional motivation in action. 'God doesn't live in a dimension where suspense is possible.'[9] It would be absurd for him to be shown striving to achieve an aim which, without a hint of strife, is already an achieved object of his all-enveloping intentionality.

By representing God in a language that discards the common grammar of intentional action and pitting against him an adversary whose language is crammed with those grammatical features God's language lacks, Milton furnishes his poem with a wide range of possibilities. Moving between the intentional activity of Satan, where each move of his plot is governed by a clearly defined prior or ultimate intention, and the intentional motive of God, where a wider, even universal sense of purposiveness renders any such precise intentions and definite aims vacuous at best, Milton inserts both modes, of existing and being, into the same narrative. As Satan attempts to incorporate God's role into his own version of events, shaping his motive, speech and action according to an alien grammar, so God envisages and in a sense contains Satan's spirit of striving, intending, willing and achieving, watching it being swallowed up in a universe that has already set a final term to all lesser and subordinate notions of purpose.

In Adam and Eve, the extreme alternatives of the erected wit of God and the infected will of Satan meet in dramatic form. Before the temptation they speak a language that lacks a strongly purposive grammar. How should they not, when what they do has the effect of realizing their own intentions only in so far as these have been preordained by God? Even in Book IX, discussion of where to do the gardening and whether to do it separately or together isn't conducted by recourse to significantly purposive forms of speech. Labouring 'to tend plant, herb and flower' (IX. 206) and choosing 'whether to wind/The woodbine round this arbour, or direct/The clasping ivy where to climb' don't sound as if they have much to do with willing – though Eve, who is speaking here, is more wilful than Adam. When Adam replies, he uses the argument that he should dissuade her from going off alone in order 'to avoid/The attempt itself, intended by our foe' (295). He associates intentional action with counter-measures taken explicitly to oppose Satan's own 'intention'. His syntax is infected by its Satanic subject.

The difference between Satan's purposive and Adam and Eve's neutral use of the infinitive is graphically demonstrated in the sequence of speeches in Book IV by Adam and Eve (411–92), Satan (505–35) and again Adam and Eve (610–88). We have already seen how Satan's speech rises on an ascending scale of purposive content. At this stage Eve is more prone to express herself in phrases such as 'God is thy law, thou mine: to know no more/Is woman's happiest knowledge and her praise' (437–8), where 'to know' is a neutral infinitive, more like Graves's 'To evoke posterity/Is to weep on your own grave' than Ulysses' 'To strive, to seek, to find, and not to yield' in Tennyson's monologue. Interestingly, though, just as Adam's syntax is infected by its references to Satan at IX. 295, so in Book IV Eve's takes on a more willed and purposive colouring when her unfallen status is psychologically most in doubt – at the pool after her birth from Adam's rib (the triple insistence on 'to look' – 458–62). And at Book V. 95–128, Adam's does the same when his explanation of their condition, apropos Eve's evil dream, is most uncertain and perplexed ('Oft in her [Reason's] absence mimic fancy wakes/To imitate her' – 110–11). Elsewhere, the only strenuously intentional speech comes when Adam tells Eve how he gave her life: 'to give thee being I lent/...to have thee by my side' (IV. 483, 485). Perhaps it is fitting that Adam's syntax should take on a pronounced Satanic colouring when he is describing the creation of Eve – both here and when it is repeated at VIII. 477–9: '... I walked/To find her, or for ever to deplore/Her loss'. But in respect of the human emotions it awakens in us, it is not Satanic at all. These are among the most moving lines in *Paradise Lost*, recalling those other words Adam spoke when he decided to eat the fruit and share Eve's life of pain and suffering. Those words were also, it will be recalled, couched in the language of purposive infinitives.

There are two important facts, then, about the language Adam and Eve speak: it is most like God's in the earlier books, before their psychology as fallen beings comes to the fore, and it is most like Satan's both when it is used to represent their wilful disregard of God's instructions from Book IX onwards and when it is used to articulate their most sympathetically human feelings (and this occurs both in their unfallen and fallen state). In one way or another Adam and Eve are sometimes God-like, sometimes Satanic. On some occasions they express God's intentional motive in words that suggest something of his own essential being. On others they copy Satan's intentional activity in a language stained with traces of the psychological motive and purposive willing that are absent from God's own speech. Unsurprisingly we encounter more of the first type of language in the earlier books, more of the second in the later ones. But they are never as uncomplicatedly one thing or the other as God's or

Satan's are. And even when they are emulating Satan's speech of strenuous purposiveness, their unfallen nature positively affects the way we respond to them. There is a difference between the relatively open-ended character of Adam's and Eve's intentions ('To live again') and Satan's ('to mar', 'to elude', 'to sit the highest', 'to avenge'). Even in their least admirable moments, Adam and Eve are not entirely dominated by too close a determination or too dark an intent.

Raised to the height of his great argument, Milton cannot avail himself of God's substitute for the intentional act, which is simply the future tense: 'What I will is fate,' where 'to will' is to acknowledge the future as a created fact. But the past tense the poet prefers is always used in the context of an appeal to his Muse, the Holy spirit: 'Say first.../... say first what cause/ Moved our grand parents' (I. 28–30). Then, within the subcontext of that past tense, he incorporates the speech acts of his characters – enunciated at varying levels of intention – out of which the bulk of the poem is made.

In the world at large, most of what Milton describes through these sub-agencies has already taken place: the rebellion, the Creation, the Fall, etc. But God's use of the future tense to indicate what we know is going to happen later in the poem, and has happened already in the world outside it, coexists with his use of the same future tense to guarantee that something is going to happen in the world outside the poem which has not already done so and which is not going to be dramatized in the main action (but will be glimpsed by Adam in Michael's vision in Books XI and XII). It is the difference between 'For man will hearken to his glozing lies.../... So will he fall/He and his faithless progeny' (God at III. 93–6), and 'Meanwhile/The world shall burn, and from her ashes spring/New Heaven and Earth, wherein the just shall dwell,/And after all their tribulations long / See golden days, fruitful of golden deeds, / With joy and love triumphing, and fair truth' (God at III. 333–8). The 'action' proceeds beyond the limits of the poem, finding its end in a condition which, in terms of the epic structure of *Paradise Lost*, is never realized, but which, in terms of God's perception of a future that is already part of the furniture of his omniscient conscience, has been realized from the outset. This knowledge of an end beyond the poem's end is Milton's too, of course – but not in the same sense as God can know it. For Milton it is genuinely and simply a fact about the future which his faith allows him to look forward to as necessarily connected to the present. This accounts for the aspect of his language which lacks intention, which eschews any sense of purposive striving.

Perhaps a better way to describe this crucial feature of the narrative of *Paradise Lost* is to return to Anscombe's sketch of an intentional act. All of her how-intentions that are ultimately swallowed up in the why-intention (of poisoning the fascists, or paying the builder) exist for God as concurrent both with one another and with their ultimate or prior intention. This is true for Milton too. So long as his language issues from the superior position his invocation to the Muse has enabled him to occupy, he is able to conceive of, and even to write about, ends and purposes which it is no part of his business to strive for or intend to realize. All of that is done for him by God's higher purposiveness, which supernaturally rises clear of the intentional acts performed on his behalf. But for much of the time, Milton allows his language to descend from this elevated position and be adopted, and therefore transformed, by lesser protagonists than God. They use it to express their motives, thoughts and actions in an altogether more intentional and purposive form. So *Paradise Lost* is full of characters striving to achieve ends which, when they are combined and expressed as a totality of all the ends towards which all the activities in the narrative aspire, must be identified with the end or purpose of the poem. But since God's intentional motive, which Milton understands and from within which he has succeeded in constructing a linguistic base, extends far beyond these merely epic or tragic ends, no purposive term is imposed on the poet's own activity. For this expresses something less narrowly intentional than anything his characters can conceive of. It is a divine condition where being and doing, ends and means, are the same and where Satan's, Mankind's, even Christ's wilful striving are no more than the poetic expression of the transcendent being of God.

DOUBLE INTENTIONS IN *TRISTRAM SHANDY*

> The night my father got me
> His mind was not on me;
> He did not plague his fancy
> To muse if I should be
> The son you see.
>
> A. E. Housman, 'The Culprit'.

Like Milton, Sterne incorporates in his narrative a model of creative power he seeks to emulate. For Milton, this was God the Creator, in whose omniscient wisdom and omnipotent control lay already the story which, for the sake of the poem, he is imagined as setting in motion. God's intentional motive is cut off from what would otherwise have been his antecedent psychological motive and consequential intentional activity. Thus God is situated in a realm of pure prior intention which causes a sequence of intentional acts only in order to render himself intelligible to fallen mankind – among whom must be numbered the readers and author of *Paradise Lost*. In making the relation of the poem to its author conform to the relation of the world to its Creator, Milton placed himself in a unique position from which to survey the end of his poem from its beginning.

How different is Sterne's relation to the world he has created. His model is not God the Father, but the merely natural father of his eponymous hero. Far from being omniscient and omnipotent, Walter Shandy is conspicuously perverse in his wisdom and limited in his power. Where Milton's God's prior intentions were no sooner conceived than they were enacted, Walter's are either nonexistent (as considered reasons for acting) or prematurely dispersed among the acts he performs in order to realize them. Nowhere is this more evident than in Tristram's conception, where the notion of premature dispersal assumes a literal, quite unmetaphorical form.

Tristram's reflections on his conception are couched in a language of strenuous though hypothetical intentionality. For him the proper way to be begotten requires that his parents should have been 'minded what they were about', 'duly considered', 'duly weighed and considered', and only then 'proceeded accordingly' (I. i. 35). He looks for a precise fit between their intention of and their intention in conceiving him, between intentional motive and the intentional act in all its logical phases. But in looking for this, he forgets about the clock, and the clock is both the accidental and the necessary accomplice in his conception.[10] It is accidental because it just so happens that Walter was in the habit of winding up the clock on the last Sunday night of every month, and that, having neglected to do this on the night in question, he was reminded by Mrs Shandy while performing his other monthly duty. It is necessary because all human actions are subject to the passage of time, and the period of time taken to perform the act of begetting is therefore fraught with possibilities of failure, interruption, incompletion and disappointment.

'Did any woman, since the creation of the world, interrupt a man with such a silly question?' Mrs Shandy might well have parried with a reminder that the creation of the world was effected by a deity who owned no clocks, let alone subjected himself to the time-consuming habit of having to wind them up. Also, his intention to invent the world was minutely realized in all the details of the Creation which so fully expressed his intentional motive. Far from this having been the case at Tristram's conception, it is unlikely that Walter intended to create his offspring by performing the act with Mrs Shandy that brought him into being. Instead he exhibited a variant of those actions of Searle's which are intentional under some descriptions, unintentional under others. We might say that Walter's begetting of Tristram, shorn of its accidental association with winding up the clock, is the equivalent of the man's sawing Mr Smith's oak plank. His intentional act of sawing a plank included the unintended accidental features of the plank's being made of oak and having come from Mr Smith's timber yard. In Tristram's case, Walter's intentional act was to perform his marital duty with Mrs Shandy. In the process, though, he performed the unintentional fortuitous acts not only of causing her to conceive a child (sawing oak), but of causing her to conceive the child that was later to turn out to be Tristram Shandy (sawing Mr Smith's oak). The fact that he did these things in the circumstances described (at half cock, as it were, because of the business with the clock) adds a further complication to an already complicated series of actions that intercalates a wide gap between a prior intention and an intentional act.

This, I wish to argue, is the model of creative activity Sterne sets before him in a novel which appears to invent the literary Tristram Shandy every bit as accidentally as Walter invents his extra-literary prototype. His passage from conception to birth, as well as from birth to maturity, is distinguished by a combination of accident, maladjustment and disarrangement. And so is the book Sterne is writing about him.

Unlike other eighteenth-century autobiographers, Tristram doesn't compose his autobiography from a point of vantage that has been occupied as a result of the activities recorded in its pages.[11] How can he, when, unlike them, he finds himself incapable of reaching that point of vantage because of something in the nature of the autobiographical enterprise that keeps holding him back? 'My work', he writes, 'is digressive, and it is progressive too, – and at the same time' (I. 27). It is true that he does try to follow the example of Defoe and anticipate that of Dickens in a number of false starts: 'I was begat in the night, betwixt the first Sunday and the first Monday in the month of March in the year of our Lord one thousand seven hundred and eighteen.... My father, you must know, who was originally a Turkey merchant, but had left off business for some years...'(I. 4); 'From this moment I am to be considered as heir apparent of the Shandy family – and it is from this point properly, that the story of my LIFE and my OPINIONS sets out . . .' (IV. 320). But it is obvious both to himself and to the least instructed reader that these are abortive and even, with their air of finality and certainty, fictitious descriptions of events. They are felt to be much less real and important than the muddle, chaos and chance occasion that surrounds them: a door closing in the present and a clock striking in the past; the author's sigh at the sadness of life at one moment and his threat to throw his pen onto the fire at the next. Then what would become of his life and opinions, or the story of them? The reality of Tristram's life and fortunes depends on their being written down in his book. But this is precisely what Tristram spends so much of his time bewailing, in writing, that he can't do. His writing is both a means and an obstacle to the recording of his life. There is something in the methods he must use to try to achieve his life in writing that prevents him from actually ever achieving it. Yet the manner of its prevention turns out to be what convinces the reader that it exists after all.

Even so, *Tristram Shandy* includes a great deal of fictional material that was conventional in its day.[12] Sometimes this is presented in the form of a story and inserted into the autobiography as a tale told by one of the characters. Such is Tristram's translation of Slawkenbergius's Tale of the long-nosed stranger in Strasbourg (Prologue to Book IV). And such would have been Corporal Trim's story of the unfortunate King of Bohemia (VIII. 19), had it not been subjected to such frequent and eventually fatal interruption. Sometimes it is a story one of the characters has shaped out of his own experience (Trim and the nursing Beguine at VIII. 22) or that of his acquaintance (Trim's intention to tell the story of Lieutenant Le Fever at V. 10 is taken over by Tristram himself at VI. 6–10; Trim starts to tell the story of his brother and the Jew's widow at II. 17, resumes it at IV. 4 and concludes it at IX. 4–7). On other occasions, a fragment of Tristram's own narrative is transformed into a self-contained story. The details of the Shandies' marriage settlement, involving the complicated legal arrangements for Mrs Shandy's confinement in London (I. 15), are of this kind. Also, one feels there are many potential stories that don't get told because of the pressure of other business. An example might be the story of aunt Dinah's elopement with the coachman (I. 21). This sounds promising, but we are not vouchsafed the details. The next and last appearance of aunt Dinah is when she dies and leaves Walter her money, just before Bobby's death at IV. 31–2. The fact that she ran off with the coachman turns out to be irrelevant.

Let us return to Trim's story of his brother and the Jew's widow, for it epitomizes the way the inset stories are related to the encompassing story of Tristram's autobiography.[13]

Trim first mentions the affair during a reading of Parson Yorick's sermon, when Uncle Toby's reference to the Inquisition recalls his brother's imprisonment (II. 17, pp. 140–1). That is to say, the end of the story is what occurs to Trim first, naturally enough, considering the event that prompted his recalling it. He refers to it over and over again as a 'story', and indeed the way Trim opens his account does sound very much like one of those inset first-person narratives from *Gargantua and Pantagruel* or the *Arabian Nights*: 'I have a poor brother who has been held fourteen years a captive...'.[14] Trim tells Uncle Toby that the story is too long to be told now, but he will tell it 'someday when I am working beside you in our fortifications'. However, 'the short of the story is this':

> That my brother Tom went over a servant to Lisbon, – and then married a Jew's widow, who kept a small shop, and sold sausages which, somehow or other, was the cause of his being taken in the middle of the night out of his bed, where he was lying with his wife and two small children, and carried directly to the Inquisition where... the poor honest lad lies confined to this hour...
> (II. 17, p. 140)

Trim becomes so overwrought at the memory that he breaks down and has to be distracted by the sermon, which he proceeds to read out to the assembled company.

The two most interesting features of the story as we have it in this first, abbreviated form are: (a) it is brought to mind because what one of the other characters says reminds Trim of its ending (the 'catastrophe' of Tom's being taken away by the Inquisition is already known); and (b) the hinge on which this inverted sequence of events turns is the reference to Tom's having been taken out of his bed because the Jew's widow sold sausages which 'somehow or other' accounted for his misfortune. Before that, the narrative moves backwards from the marriage to the sausage making; afterwards, it moves forwards to the Inquisition, where we already know it will end. The crux of the story, then, is the connection between the sausage making and the night arrest. This is the only aspect of the plot that we are not told about in the first account, and we are not told about it in the second (brief) account either. This merely repeats the fact that Tom's misfortune arose from 'marrying a Jew's widow who sold sausages' (IV. 4, p. 277). The upshot is that we expect the link between the sausage making and the arrest to figure prominently, 'from first to last', in the third account.

Incidentally, the sermon, which acted as a sort of frame to the first, abridged edition of the story, is provided with a much more straightforward history than Trim's brother ever acquires:

> Ill fated sermon! Thou wast lost, after this recovery of thee, a second time, dropped through an unsuspected fissure in thy master's pocket, down into a treacherous and tattered lining – trod deep into the dirt by the left hand foot of his Rosinante, inhumanly stepping upon thee as thou falledst; – buried ten days in the mire raised up out of it by a beggar, sold for a halfpenny to a parish-clerk, transferred to his parson, – lost for ever to thine own, the remainder of his days, – now restored to his restless MANES till this very moment, that I tell the world the story.
> (II. 17, p. 157)

The story of the sermon is as much superior to the story of Trim's brother as the story of Tristram as a homunculus in I. 2 is superior to the story of Tristram after he has been born. The homunculus, it will be remembered, was a 'young traveller', a 'little gentleman', who struggles with melancholy dreams and fancies through his nine-month journey until, at last, 'his muscular strength and virility worn down to a thread', he completes his history and reaches his 'journey's end'. Like the sermon's, this is a complete narrative told from beginning to end. The sermon and the homunculus are both picaresque heroes suffering the checks and adversities common to their kind before achieving a comic apotheosis in the restoration to Yorick's MANES and the completion of the journey. Nothing like this happens to Tristram at the centre of the story, or to Trim's brother at its periphery. It seems that only sheets of paper and homunculi are suitable heroes of tales that are 'properly' told.

Tom's story is resumed at IX. 4–7, at a point in the outer narrative where Toby and Corporal Trim are on their way to the Widow Wadman's house to make her a proposal of marriage. These are not the circumstances in which Trim had intended to tell Toby the full story, 'some day when I am working beside you in our fortifications'. Such a day, described in Book VIII, was taken up with Trim's abortive attempt to tell him about the King of Bohemia and the more successful narrative of his affair with the Beguine ('the essence of all love romances'). At present, Toby and Trim are engaged in a more dramatically purposive activity than staging the siege of Namur on Uncle Toby's bowling green. It is a climactic incident. The widow has prepared her stratagem with Bridget, Walter has written Toby his letter about love-making, and Toby has steeled himself to the prospect of making his proposal of marriage. Walking to the widow's door with the intention of proposing in mind, Toby and Trim are spied upon by Walter and Mrs Shandy from an angle of the old garden wall. The whole story is told while master and servant are standing in front of the door, framed by the observation posts of the Shandys (who are wondering 'what can their two noddles be about?') and propelled there by what seems to have been an infinite series of causes, plots and accidents. The form of the narrative is heavily determined by the function it has to perform – i.e. the dramatization of a climactic event in the story of Toby and the Widow Wadman, which is the principal subject of the last two books of the novel. But instead of releasing the event, the action pauses for four chapters while Trim resumes his story of Tom and the Jew's widow.

Why does Tristram choose to tell the third and final version of the story at this particular moment and in these particular circumstances? There appear to be three reasons.

The first is what we might call the obviously plausible psychological reason. Toby, on arrival at the door, is nervous about whether the widow will accept his proposal. 'She will take it, an' please your honour, said the corporal, just as the Jew's widow at Lisbon took it of my brother Tom.' That is to say, Trim tries to boost Toby's confidence by telling him a story that includes a successful proposal of marriage. That this is not just an excuse to tell the story is shown by the way it is told: Trim passes over parts of the narrative that are manifestly important, so as to emphasize the 'proposal scene'. In fact, it is with this scene that Trim ends the story – Tom having won the widow's hand by means of some crafty (and possibly bawdy) foreplay with the sausages: 'She signed the capitulation – and Tom sealed it; and there was an end of the matter'. But for us, there is only the beginning of the matter. What we want to know about is how the business with the sausages led to Tom's imprisonment by the Inquisition.

The second reason is that Trim recognizes Toby is unwilling to go through with the proposal immediately and he needs to play for time. Something has to be done while they wait on the doorstep. So he tells Toby a story. We might call this the not-so-obviously psychologically plausible reason, because we are likely to feel that Trim's decision to tell a story here is as much for the convenience of the author as it is for the benefit of either of the characters. There is a pause in the narrative during which nothing happens. How is the novelist to keep the reader occupied during such a pause? By telling a story. But he has to tell it through the mouth of one of the characters because there has to be something for Walter and Mrs Shandy to see and discuss (i.e. in this instance to mistake the telling of story for a discussion about fortifications). This would explain some other features of the story Trim tells – other, that is to say, than the emphasis that falls on the proposal. But since these are even more plausibly accounted for by Tristram's third reason for telling this version of the story, I shall defer discussion of it until after I have explained what I take this to be.

The third reason is that Tristram/Sterne knows he is approaching the end of his 'Life and Opinions' and he wants to take up the story of Trim and the Jew's widow that was left dangling back in Books II and IV. The psychological plausibility attaches to the author, not to the characters, and, unlike theirs, it is not connected to any idea of a specific purpose that the telling of the story is intended to fulfil. Sterne behaves as Searle did when he suddenly hit the man. He had no prior intention of doing so, he just did it – i.e. Sterne just told it. But because he was aware of the fact that he was shadow-writing Tristram's autobiography, he had to give it to Trim to tell, and because in novels and autobiographies people don't tell stories just because they want to tell them, he placed Trim in a situation that could be construed as providing a reason for doing it. The aspect of the narrative that will be most likely to impress us is not so much its ending with the proposal scene as its ending with a scene that is not the end of the story. Sterne can't resist not telling us the whole story even though his decision to resume and repeat it sprang from his recognition that he was getting close to the end of the longer story in which he intended to include it.[15]

In this third version of the story we still don't have an answer to the most pressing question that the first and second versions raised in our minds: how was the cause ('somehow or other') of Tom's being taken in the night by the Inquisition related to the fact that the widow was a sausage maker? We are not told in so many words. It is merely hinted that the answer has to do with the fact that the widow was Jewish. This, at any rate, I take to be the point of the opening complaint that Tom would not have endured such misfortune 'had it pleased God after their marriage, that they had but put pork into their sausages'. Then 'the honest soul had never been...dragged to the inquisition'. But it is all very circumstantial and enigmatic, with no details given about what must have been the religious authorities' change of attitude to Jewish sausage makers.

This lack of detail about a crucial aspect of the story is all the more remarkable when placed beside the extraordinary excess of detail in other, less significant episodes. Look, for instance, at the treatment of the poor negro girl who was the only person in the sausage shop when Tom arrived with his proposal. Although she appears to have no role to play in the drama, we are told that she had 'a bunch of white feathers slightly tied to the end of a long cane, flapping away flies – not killing them'. There are two questions that spring to mind about the negro girl. One is the general question as to why she is there at all, doing something very specific that has no bearing on the story. The other is why she doesn't kill the flies, but is described, scrupulously, as merely 'flapping' them away.

The answer to the first question might be that at some later stage in the story we haven't yet heard about, this negro girl is going to play an important part. After telling us about the cane, Trim says that 'there are circumstances in the story of that poor friendless slut, that would melt a heart of stone . . .; and some dismal winter's evening when your honour is in the humour, they shall be told you with the rest of Tom's story, for it makes a part of it –'. So, she is more important than at first she seems. But what possible role can she have to play? When we moved from the first version to the second, we thought that the only part of the plot that needed explaining was the reason for the Inquisition's persecuting Tom – which had something to do with the sausages. Now that we are moving towards the end of the second version, we find that a completely fresh complication has arisen. That is why Trim begins the next chapter with the words: 'As Tom...had no business *at that time* [my italics] with the Moorish [sic] girl...' No sooner is one mystery (half-)solved, than another mystery pops up. The more that is explained about a story, the more there will be left to explain – a not unfamiliar problem in Tristram's autobiography. It goes some way towards explaining why Tristram/Sterne wanted to have another try at telling the story before the autobiography was finished. But this propensity for stories to generate other stories might also be interpreted as having a bearing on Trim's not so obviously plausible psychological reason for telling the story in front of the Widow Wadman's door. By playing out the narrative at such length and incorporating such arbitrary details in it, he is succeeding in his less specific purpose of playing for time while Toby regains his confidence. So it is possible to explain this aspect of the story in terms of the requirements of either the characters or the story teller, or both of them if our critical demands are sufficiently elastic.

The second question raised by the presence of the negro girl has to be answered by acknowledging that the demands of the autobiographer are surplus to the requirements of the tale teller. We might be able to relate the girl's kindness to the flies to the psychological attitudes and motives of the character who remarks on it. But the principal application is to the novelist, not to Trim.

The main point about the reference to flies is that it brings to mind another place in the novel – nothing to do with the story of Trim's brother – where flies, and kindness to flies, play an important role. This is at II. 12 (pp. 130–1), after Dr Slop's arrival at Mrs Shandy's lying in, where Tristram recalls an incident expressive of Uncle Toby's kindly disposition. Toby, he says, 'had scarce a heart to retaliate upon a fly' and describes how he caught a fly that was buzzing around his nose, went and opened a window, and let it out of the room. The action, says Tristram, 'instantly set my whole frame into one vibration of the most pleasurable sensation'. It is a passage justly celebrated for expressing Tristram's sentimental attitude. Is it not more likely, then, that at the Widow Wadman's door in Book IX, Sterne is allowing the course of Tristram's narrative to be influenced by suggestions thrust forward by memory of an earlier incident described in the outer 'casing' of the Tristram/Walter/Toby story than that Corporal Trim is allowing his narrative to be influenced by notions of what he supposes is psychologically appropriate to his fly-loving master's predicament in the inset story of Tom's marriage to the sausage maker? There is room for both explanations. But here I think the balance of probability inclines more towards Tristram's than Trim's intentions, whereas in the matter of the amount of time (Trim) and space (Tristram) given over to the description of the negro girl the balance is pretty well even.

Three intentional acts are contained one inside the other in the third version of the narrative of Trim's brother. There is the strongly motivated narrative serving a specific purpose in Trim's telling the story to increase Toby's confidence. There is the equally strongly motivated narrative, but with a more ambiguous and less specific purpose, in Trim's telling the story on Tristram's behalf in order to pass the time while Toby and Trim stand waiting on the doorstep. And there is the weakly motivated narrative (if motive is considered as something other and more precise than 'wanting to tell a story just because you want to') serving no purpose beyond empowering Tristram, whose brain-child it obviously is, to pull in, and then tease out again, a loose narrative thread he wants to remind us about before racing forward to the end of the book.

No wonder the 'sudden transitions' caused Trim to lose 'the sportable key of his voice which gave sense and spirit to his tale' as he 'attempted twice to resume it, but could not please himself'. And no wonder that, reflecting on the story from the Shandies' position outside it, Tristram says that it 'went on – and on – it had episodes in it – the reader found it very long –'. In order to write this, Tristram must have shaken himself out of his narrative daydream – which has a very present feel to it when it is being heard – and set it at a distance from himself, as a 'story' that has by now been written down and read, very much in the past tense. But at the time it was being reported, in Trim's narrative, it made the opposite impression. It might have been long, episodic, apparently endless – but by virtue of the fact that it could have gone anywhere and spread out in all directions. It had no end-directed shape. Tristram's intentions in telling it through Trim, whatever Trim's intention was of telling it to Toby, seemed to be governed by no conscious prior intention on Sterne's part related to any other purpose than that of filling up the space to his own satisfaction.

In the story of Trim's brother and the Jew's widow, the motive of the character who tells it is overlaid by the absence of motive of the author who has created the character. The author slides in and out of the character's point of view in order to narrow or widen the scope of his purposive activity in writing about it. The story ends where what is most in evidence from the vantage point of the characters' behaviour is the description of an event that we expect will justify Trim's motives in telling it – the successful proposal. The overwhelming impression the whole story makes on us, however, is very different. It is one of a multitude of possibilities prematurely foreclosed, but, we suspect (with the earlier version of the story to guide us), likely to spring open again at any moment – unless *Tristram Shandy* itself manages to stop. If that doesn't happen, the opportunities are endless. But by virtue of being endless, they are pointless too. Purposeless, in fact. And the same thing is true when the current of the story flows in the opposite direction, when we are encouraged to work back from the ending of a story to its beginning. This way round, the ending is determinate, but the beginning is indistinct. The river that ends its course at a single mouth is fed by a host of tributaries, each one of them issuing from a mysterious origin in invisible springs and sources.

Although the reader of *Tristram Shandy* spends a good deal of time waiting on events, when they arrive they often do so in a precipitate, even headlong way. The first version of the story of Tom and the Jew's widow began with Tom's arrest by the Inquisition. This was a consequence of Trim's reading of Yorick's sermon, which suddenly carried us to the conclusion of the story before its beginning was ever hinted at. In the same way, events rush upon us in the main business of the autobiography. No sooner are we told that 'if an event had not happened' then 'I verily believe' such and such would have ensued, than it seems to burst upon us out of nowhere. In the passage referred to, Walter Shandy has for some time been engrossed in the writing of his *Tristra-paedia*, and it seems that we are to prepare ourselves for a lengthy account of his progress. Judging by the time it has taken to get Tristram born (two hundred and fifty pages in the Penguin edition), we may be surprised to find three years' composition of the *Tristra-paedia* passing by in a single paragraph, and suppose they will be reviewed later at more leisure and with less speed. However, the impetus is suddenly transferred from Walter's writing to Tristram's living, and it looks as if the pace of the narrative will quicken accordingly. We are told that Walter advanced very slowly with his work. But the fact that Tristram 'began to live and get forwards at such a rate' introduces a note of desperation. The writing and the life seem to be tripping over themselves in an effort to keep up with each other. This is Walter's literary problem in a nutshell, and Tristram's also, in so far as he too is a writer, as well as a liver, of his life. (See IV. 13: 'I shall never be able to overtake myself.')

Then, suddenly, the event referred to takes place: 'twas nothing'. It is over before we have had time to recognize it. Opinions about it rush in thick and fast, far in advance of our being able to determine what, exactly, has happened: 'I did not lose two drops of blood by it – 'twas not worth calling in a surgeon, had he lived next door to us –'. We are puzzled by the confession that 'thousands suffer by choice, what I did by accident', and misled by what appears to be, in the inadequate context in which we read it, a metaphorical description of a moral act: 'some men rise, by the art of hanging great weights upon small wires.' In spite of the reference to sash windows further down the same paragraph, we have to wait another two chapters before the nature of the accident, and with it the meaning of the aphorism, is made clear to us. Tristram has been injured because the sash fell on his prick as he was peeing out of the window. Such is the information given in the first two paragraphs of V. 17, pp. 369–75. That is the event, and in spite of the confusion of the telling, it is all over and done with in less than half a page. But just as Tom's journey to Lisbon was only the start of a narrative that, as it moved forward in time, became ever more dispersed and inconclusive in content, so Tristram's accident is only the conclusion of a narrative that, as it moves backwards in time, becomes ever more dispersed and inconspicuous in its causes. Where the effects of Tom's liaison with the widow race far ahead of Trim's capacity to catch up with them all in his story, the causes of Trim's accident fall back too deeply into the past for him to dredge them all up to the surface of his history. Whether we move forward from cause to effect, or backwards from effect to cause, the further we get, it seems, the further we have to go.

How and why was Tristram circumcised (as it later transpires) when he was five years old? In the order of telling, we would have to say that it happened because the chambermaid forgot to leave a piss-pot under the bed, with the result that Susannah encouraged him to pee out of the window, the sash fell down and the damage was done. But Tristram is not satisfied with this explanation. He foregoes the opportunity of investigating why the chamber-maid forgot but feels he must supply an explanation as to how it was that the sash fell down. Presumably he feels he must do this because Susannah fled for protection to Uncle Toby's house, and Trim, who lived with him there, happened to be the obvious culprit. For Trim had taken the two lead weights from the nursery window in order to provide Toby with a couple of field pieces to mount in the gorge of his new redoubt. (Details are given as to why he couldn't use the weight from Toby's own sash windows – already dismantled, along with much of the rest of the house, to produce eight new battering cannons and three demi-culverins.) So the cause of Tristram's accident really lies in Toby's mania for war games, and to find an explanation for that we have to go back to the siege of Namur and beyond.... Quite apart from this, it is hinted as well that Walter Shandy might be to blame because, although he had not yet written his chapter on sash windows for the *Tristra-paedia*, he appears to have been fully apprised of the danger they represented: 'Obadiah [was] enabled to give him [Walter] a particular account of it, just as it happened – I thought as much, said my father, tucking up his night-gown; and so walked upstairs.' He walks up the stairs for reasons that remain mysterious (is the 'statutable minute' Tristram mentions the time he takes inspecting the wound?), and the next time we see him he is walking down again – not to fetch the lint and basilicon required for a dressing but to consult 'a couple of folios' on the origins and purposes of circumcision.

The search for the reasons for Tristram's accident has moved rapidly from the absence of a chamber pot back into the mists of biblical and pre-biblical time. In the end, it might all have to do with fate and the stars. As Walter says, 'There has been certainly... the deuce and all to do in some part or other of the ecliptic, when this offspring of mine was formed.' And the matter of Tristram's circumcision disappears into the endless polemic and controversy of Walter's conversation with Yorick and Uncle Toby. In the development of the plot, it has the effect of persuading Walter to take Tristram out of the hands of the women of the house and place him under the supervision of a tutor. Hence the introduction of Le Fever's son into the story. But the son's role (as tutor) is subordinated to that of his father in Toby's story about him, and in fact he never makes a contribution to Tristram's narrative. The circumcision seems to make no difference to Tristram's fortunes. The pressure it exerts on the narrative is not forward to the description of effects, but backwards to the elucidation of causes. In this respect it anticipates Poe's construction of 'The Raven', according to the author's account of it in 'The philosophy of composition': a movement from a known conclusion to a determinable origin.[16] For the Shandy household, though, there is no determinable origin – unless it be in the ritual practices of the Hebrews and Egyptians or the motion of the stars. Even if there were such a termination, it would not be a single one, the stopping place of a movement backwards in one direction. What of the other directions the narrative might have taken to account for this sorry episode: inquiring into the chambermaid's neglect? or Susannah's contrivance? or Trim's and Toby's war games? The regressive potential of the incident is limitless. As with Tom and the Jew's widow, only the selective and associative inclinations of the author, and the physical constraints of the book itself, serve to bring it within bounds.

In both these episodes Sterne's inclination to propel his narrative forward, by providing causal explanations or creating the impression of psychological plausibility, is transformed into something seemingly detached from any governing intention. It has taken on a life of its own. The intentional act of writing seems to have pulled loose from any (inferred) prior intention more definite than that of writing about the life and opinions of Tristram. Consequently the intention in writing one part of the novel may be related only tenuously to the intention in writing another part of it, often more in terms of random association than of a coherent development of the narrative. This method of writing is fortified by Sterne's interest in associationist psychology.[17] But really that is only the gilt on the gingerbread of his narrative method which, though it appears in his work in a singularly pure state, is in fact the same as that of most successful novels, and other forms of story-telling too.

For in these two episodes, as in the novel as a whole, Tristram contributes to the achievement as fiction of what his father has achieved in fact: namely his own coming into existence. Walter might not have intended to beget a child on that fateful Sunday night, but having begotten him, he certainly intends to exert a controlling influence over his gestation, birth, upbringing and education. The results, as we see, are very different from those he intended. Tristram on the other hand does most certainly intend to write the history of his life and opinions. But that is a very general aim indeed. When he comes to tell the details of how he was born, brought up, educated and settled in life (if he ever is), the way he goes about the telling has the effect of hastening him far beyond anything that could reasonably be construed as having been a conscious part of his prior intention. Just as Walter's animal spirits were dispersed in a bizarre combination of biological necessity and environmental hazard, so are Tristram's autobiographical intentions dissolved in a mysterious chemistry of conversations by turns desultory and frenetic, and unfinished (unfinishable) stories and anecdotes. And just as Walter's subsequent efforts to control the course of Tristram's passage through life are thwarted by a cut thumb, a pair of tight breeches, a hot chestnut and a faulty sash window, so Tristram's efforts to control the narrative of his passage through life are thwarted by distracting puns, pages lost, sheets of manuscript accidentally thrown into the fire, sermons that slip out of treatises on engineering and mysterious discoveries in the margins of the sermons themselves.

The accidental though inevitable frustration of Walter's intention of bringing up his second son, baptized Trismegistus, with a sharp nose and an uncircumcised penis, and properly educating him through the writing of the *Tristra-paedia*, is matched by the equally accidental and equally inevitable frustration of Tristram's intention of representing his life and opinions in the manner prescribed by Sterne's immediate predecessors in the writing of fiction: i.e. 'I was born at Shandy Hall in the county of Yorkshire, was educated at home by my father Walter Shandy and private tutor Mr Le Fever, became heir to the property on my brother Bobby's death, went on the grand tour at the age of... *etc*. In consequence I have formed the following opinions...' Instead of this, Walter's son is baptized Tristram. He has a nose like the ace of clubs, a circumcised penis and no education to speak of because of the difficulty his father experiences in composing the *Tristra-paedia* fast enough for him to be able to read it in time. Tristram's autobiographical subject suffers from the same deficiency as the *Tristra-paedia*. He too cannot keep pace with his author's efforts to provide him with a history. It is difficult enough for Tristram to bring himself to birth at all: he doesn't succeed in doing so until about half way through the book. Even then, the record of his later history is continually deflected by conversations, disquisitions and stories that often have the effect of returning us to a time before he was born. The last two books, for example, describe events that were over several years before Tristram's arrival on the scene.

These peculiarities are aspects of the novel's surface, not hidden mysteries to be exposed only by the most scrupulous investigation of its depths. Sterne is entirely frank about the coincidence of the events he is writing about and the event of his writing. How otherwise explain his description of Walter's reaction to the news that his second son has been christened Tristram, in spite of all the precautions he has taken against it? He compares Walter's disappointment now with his earlier disappointment when Dr Slop disfigured Tristram's nose with his forceps. Then, his reaction was to take to his bed and speak to nobody. Now, he reacts quite differently to this new mortification:

> The different weight, dear Sir, – nay even the different package of two vexations of the same weight – make a very wide difference in our manner of bearing and getting through with them. – It is not half an hour ago, when (in the great hurry and precipitation of a poor devil's writing for daily bread) I threw a fair sheet, which I had just finished, and carefully wrote out, slap into the fire, instead of the foul one.
>
> Instantly I snatched off my wig, and threw it perpendicularly, with all imaginable violence, up to the top of the room – indeed I caught it as it fell – but there was an end of the matter; nor do I think any thing else in Nature, would have given such immediate ease....
>
> Now, my father could not lie down with this affliction for his life – nor could he carry it upstairs like the other – he walked composedly out with it to the fish-pond...: there is something, Sir, in fish-ponds...
>
> (IV. 17, pp. 291–2)

He tells us that he has no idea of what it is. But it works. Just as Tristram's own spontaneous gesture of throwing his wig up into the air worked. Tristram confesses to throwing his wig in the air in order to add conviction to his description of Walter's walk to the fish-pond – as if mentioning his walk to the fish-pond reminded him of throwing his wig in the air. But we might prefer to read it the other way round, noticing that he threw his wig in the air less than half an hour ago. Surely it may be that the memory of throwing his wig in the air prompts him now to use the equally irrational device of Walter's walk to the fish-pond as a plausible reaction to his disappointment over the name. Whichever way we look, it is hard to disentangle Sterne's behaviour in writing the novel from Tristram's and the other characters' behaviour in living between its covers.[18]

MILTON AND STERNE

> Lord, since eternity is Thine, art Thou ignorant of what I say to Thee? or dost Thou see in time, what passeth in time?
> St Augustine.

In recent years many novelists have gone to school with Sterne. The tradition extends from Beckett's *Unnamable*, through Robbe-Grillet's soldier with the mysterious parcel in *Dans le labyrinthe*, to John Barth's self-conscious novelist in *Lost in the Funhouse* and B. S. Johnson's mendacious architect in *Albert Angelo*. What makes Sterne so different from these authors is the genial spirit with which he tempers what might otherwise have been the desperation of his language. Near the beginning of *The Unnamable*, Beckett's monologist asks himself why he is committing his thought to speech, or to paper, at all: 'One starts things moving without a thought of how to stop them. In order to speak. One starts speaking as if it were possible to stop at will. It is better so. The search for the means to put an end to things, an end to speech, is what enables the discourse to continue.'[19] For the Unnamable, speech is the only means he can find of putting an end to speech. It is an inevitable fact about the human condition. There is no joy in it, no creative satisfaction. It is just there, once started, and there is no hope of not starting it. It won't stop until there is 'an end of things' altogether – which seems desirable, given the pointlessness of speech. But it has no end in the sense of purpose, point, reason for being there. In other words, it articulates no purposeful intention. There is no question of it achieving anything other than its own insignificant existence – now in the author's mind, now on the page, now in the reader's mind. It communicates nothing.

What is it in *Tristram Shandy* that, while not being in any sense a search for the means to 'put an end to things', nevertheless 'enables the discourse to continue'? I would argue that it is Sterne's intransitive intention, transferred by proxy to Tristram, to tell Tristram's story. It has to be expressed as generally as that, in order to prevent the sense of purposive activity from attaching itself to too specific an aim, dominated by the too narrowly motivated expression of the author's prior intention. But we have to suppose that Sterne is genuinely interested in telling Tristram's story, and therefore Tristram is genuinely interested in telling his own story. But in making the attempt he is operating at the furthest remove possible from Milton telling his story of the Fall of the Angels and the Fall of Man. For Milton isn't telling a story about a group of characters in whom he just happens to have taken an interest. He is telling *the* story of mankind, which is all-inclusive, definite and, above all, over. Over, that is to say, from God's point of view – which is *a* point of view the poet is at liberty to *try* to adopt at the same time as it is *the* point of view, and the *only* point of view, from which all of the true story can be observed.

Milton couldn't possibly place himself in a position from which he could see things from God's point of view. Knowing this, he invented a stratagem to draw attention to the limitations inherent in his attempt to tell this particular story, and by doing so obliquely suggested what the real shape of the story might be. In *Paradise Lost* he interpreted God's creation of the world as the ultimate type of the creative act which, to an infinitesimally smaller degree, he copied in the act of writing his poem. But he wrote *Paradise Lost* about God's creation too. So God's activity, speech and vision became both the model Milton aspired to emulate and the subject on behalf of which he sought to emulate it.

Milton knew this was impossible. It would be absurd for anyone to have planned to write a poem, the purpose of which was to explain the Fall of Man from that same God-like position from which he must acknowledge Man has fallen. What he can do, though, is to create a very generalized picture of God, and then make his fictional presence more vivid by contrast with other characters in the story whose peculiarities, especially the peculiarities of their language, are in such marked contrast with his own. He succeeded in this by emphasizing and exaggerating different levels of intentional activity in the speeches of the characters, showing how they strain to achieve ends which we know God has already brought into existence simply by virtue of his being what he is – omnipotent, omniscient and, unlike Walter Shandy, untroubled by the presence in his life of wives, clocks and calendars. *Paradise Lost* interests us as a poem because it plays off the low intentional activity of the lesser beings who populate it against the pure intentionality of God.

Pure intentionality is that aspect of God which accounts for his being able to have intentions without needing to realize them as intentional acts. It is a paradox that because God is omnipotent, all-powerful, he can do anything, but because he is also omniscient, all-knowing, he won't do it – because it is already done. If something is known to be, it is. There is no need to bring it into being. Or, put another way: for God, to know something is to have already created it, although at no point in the 'past' was it created before it was known. Milton would have been familiar with the problem from his reading of St Augustine:

> His substance is no ways changed by time, nor His will separate from His substance. Wherefore He willeth not one thing now, another anon, but once, and at once, and always. He willeth all things that He willeth; not again and again, nor now this, now that; nor willeth afterwards, what before He willed not, nor willeth not, what before He willed; because such a will is mutable; and no mutable thing is eternal; but our God is eternal.[20]

This notion of God's essential being would be naive theology were it not for the fact that it has such startling literary consequences. For the world God holds suspended in his pure intentionality is one whose inhabitants do all the intending, designing, being purposeful, forming plans, *etc.* that God doesn't need to do. For the poet who seeks to emulate God in this respect, the plot of his poem lies open to the view as a sort of object in space that can be inspected from many angles or seen all at once – as God sees his creation. 'For what was spoken was not spoken successively, one thing concluded that the next might be spoken, but all things together and eternally. Else have we time and change; and not true eternity nor true immortality.'[21] What is the same in God's and Milton's accounts of the Fall is the relation each Fall bears to its creator. Narrative of the Fall is to Milton as poet, as Fall itself is to God as Creator. But that is not to say that the Fall itself is not a story. Or at any rate that is what it becomes when recounted by any human agency: the author of the Book of Genesis, the French poet Du Bartas or Milton himself. But because Milton can take God's Creation of the world as a model for his own infinitely less impressive creation of the story of the Creation of the world in *Paradise Lost*, his attitude towards the characters in his poem can also have something in common with God's attitude to the real beings in his world.

If Milton had maintained this lofty attitude throughout *Paradise Lost*, he would have written a very dull story. God's purposive purposelessness, his pure intentionality, transcends our capacity to be interested or entertained because it transcends the human fact of time. But poems can only live in time and, rhythmically and formally, by time. So Milton's emulation of God's relaxed, unpurposive sheer being has to remain for the most part merely a possibility, a potential state from which the poet can, and must, descend for the greater part of the duration of his narrative. In fact the only occasions on which he assumes the poetic equivalents of the visionary and linguistic attributes of God are those on which he seeks to describe God directly or allows him to speak directly to an audience within the poem. This he does very seldom, for poetic reasons which will be immediately apparent. In much the greater part of the poem, Milton is engaged in the very different task of imitating the more or less actively intentional states of the other characters.

When Milton is recording for his audience the activities of characters such as Adam, Eve, Satan or even Christ, he is presenting something at a level of psychological motive and purposive intention that, from the simulated point of view of God, he knows to be already 'accounted for' in the wider area of God's intentional motive. While for Macbeth the murder of Duncan is effected by setting in motion the intentional act of killing the king – 'I go, and it is done' – for God in *Paradise Lost* the whole scheme of universal history is realized merely through its being known. 'I *know*, and it is done,' God might have said in a more laconic and less impressive poem. But Milton can only really 'know' in the human sense of that word. So he builds up his story for the most part in terms of his limited knowledge of human motive and psychology. Nevertheless, the validating presence of God's pure intentionality accounts, at a level strictly speaking unrealizable in the poem, for everything that Milton describes to us in terms of mere intention, mere purpose. Milton's knowledge that it is there, built into the language and structure of his poem, is the ultimate source of the extreme contrast between his work and Sterne's.

Sterne occupies a diametrically opposite position to Milton's. Far from having as his model a God for whom the created world exists wholly in intention before it is 'realized' in a sequence of intentional acts, Sterne has no better model than his hero's father. All the others immediately to hand have proved useless almost from the start. Hence the collapse in the first three chapters of Tristram's feeble attempt to produce a story of his life in the manner of Defoe or Fielding. No sooner does he put pen to paper than the intractability of his problem and the defeat of his purpose are shown to be equally inevitable. Negatively, he has the model of his father's *Tristra-paedia* to warn him. As indeed it does, for his comments on it are as relevant to his own doomed enterprise as they are to Sir Walter's: '– Certainly it was ordained', he writes, 'as a scourge upon the pride of human wisdom, That the wisest of us all should thus outwit ourselves, and eternally forego our purposes in the intemperate act of pursuing them' (V. 16, p. 369). But Tristram's and Sterne's insistence on pursuing these purposes in spite of Sir Walter's warning has a positive consequence all the same. Purposes foregone can produce alternative satisfactions in the very processes of their author's foregoing them.[22] In *Tristram Shandy* this is what they do, though Sterne seems to be trying to prevent them. Narratives seem to require a high motivational content in their authors, transmitted to their characters in the form of purposive aims and intentional acts, not in order that any one of these shall be appropriately realized in the development of the story but in order that the opposite of this shall take place. Intentions in action are generated by a prior intention, and are designed to bring about the most direct and immediate satisfaction of that prior intention through its achieving its intentional object. Typically the author will have a prior intention that he seeks to satisfy through the agency of intentions in action which generate as little purposive activity of their own as is consistent with their identity as intentions in action.

Only God can perfectly achieve this. The author discovers how un-Godlike he is when he realizes the extent to which these intentions in action slip away from his control. But they rarely slip away from his influence. The way the sense of purposiveness is achieved at the same time as the specific purposes are 'foregone' is by the persistent presence, at some level of textual activity, of authorial influence retained where direct authorial control has been lost.

Between these two extremes most stories manage to get themselves told. Occasions on which writers, like Milton, usurp the intentional motive of God are as rare as those on which others, like Sterne, rely so entirely on their own expertise. Other narrative poems than Milton's have been more than the sum of their parts, but very few indeed have contrived to precede that sum. Nevertheless the reader detects in some stories more than in others the stable presence of a less metaphysically grounded prior intention, to which the intentions in action of the incidents and episodes to a greater or lesser extent conform. Sometimes this is signalled by a preference for certain formal arrangements over others: third-person omniscient narrative over first-person narrative or some other variant of the indirect or 'interested' approach. But more often than not the search for equivalence between structural features and degrees of prior intentionality will be merely misleading. An interesting secular example of a narrative in which all is 'known' by the author beforehand, and therefore the control exerted over the telling by his prior intention is very considerable, is Garcia Marquez's *Chronicle of a Death Foretold*. Here we have a fascinating conflation of Miltonic omniscience and Sternean movement from effects to causes, which produces a story with a boldly emphatic outline and a mysteriouly complex narrative 'mass' around which it has been drawn. There is no straightforward connection to be made between the third-person narrative form in which the story is told and the degrees of knowledge and ignorance readers possess at different stages in its telling. We are dealing with intangible qualities that rarely proclaim themselves in the technical point of view the author adopts towards the activities he describes.

In real life the realization of ultimate intentions invariably has to compete with the characteristic propensity of intentions in action to create a momentum of their own which, even where it doesn't affect the plot (which it usually does), is bound to affect the narrative at lower levels of expressive or mimetic activity. Nevertheless, it is obvious that novels like *War and Peace, The Human Predicament* and *The Raj Quartet* occupy a position somewhere at the Miltonic end of the intentional spectrum. At the 'lower' levels of scene, description, episode and plot, we sense a conformity with a sense of purpose that is the manifestation of a governing intention. In novels by Dickens, James and Dostoevsky, on the other hand, though evidence of a governing intention sometimes makes itself felt (in Dostoevsky, for example, in an almost programmatic form), the life of the narrative evidently springs from intentions in action that create provisional and changeable purposes according to which the success or failure of their representation is to be judged. In both kinds of narrative, the satisfaction the reader feels will be strictly in proportion to the sense of inevitability he acquires in the novel's, play's, poem's having become what it has become at its close. But the relative importance of prediction and retrospective surprise in the face of that sense of inevitability will differ from the one kind of narrative to the other. At the beginning of *Paradise Lost* we look forward, with God or the Holy Spirit, to a completed action that the rest of the narrative will re-present. At the end of *Tristram Shandy* we look back with a baffled and bewildered autobiographer to a completed story that seems to have scarcely begun to give an account of its subject.

SPEAKING MEANING IN 'AN EPISTLE TO A PATRON'

En tout chose il faut considérer la fin.
La Fontaine, 'The Fox and the Goat'.

To discuss, or even think about, these different types of success and failure without recourse to judgements about authorial intentions seems to me self-defeating if not impossible. I am aware, though, that in placing the emphasis on dramatic or other forms of narrative fiction I might be making things easy for myself. After all, in these texts the author is able to make use of his characters as objects of his attention and as devices to publicize his intentional motives in intentional acts relating to each other in significant ways. How they do so, at different levels of conscious and purposive deliberation, has been the subject of the preceding pages.

What happens, though, in the analysis of non-dramatic poetry, where no such characters exist to mobilize their authors' intentions? How do we arrive at a description of a lyric poem, for instance, in terms of its author's realization of his intentional motives or organization of his intentions in action? Such poems rarely contain characters. And when they do, their motives, intentions and purposes are unlikely to be represented as being of any significant interest to the reader. Who ever dreamed of considering Keats's gleaner or Hopkins' ploughman in this light, even after their political significance has been taken account of? Of course, if a human being is the subject of a lyric poem – Wordsworth's solitary reaper, for instance – the presence or absence of a sense of purpose in what he does might well become an important issue, but not one that will help deliver the poem to us in a proper, formal sense. Intentions in poems such as these display themselves in different ways, and guide us in the act of reading by performing very different functions from those undertaken by characters in stories.

I want to examine how they do this by looking in detail at a poem that is as little dependent as it is possible to be on notions of character, ideology, subject or any kind of what might loosely be described as extra-poetic 'meaning'. For this purpose I have chosen a little-known, but in my opinion excellent, poem by the South African poet F. T. Prince. It is called 'An Epistle to a Patron' and can be found at the opening of Prince's first volume, *Poems* (1938), of his first collection, *The Doors of Stone* (1963), or of the *Collected Poems* (1979).[23]

The only reference I have been able to find to 'An Epistle to a Patron' in a readily available work of criticism is in Donald Davie's *Articulate Energy*,[24] where it is described as a poem that is 'pure', 'absolute'. 'The poem does not even explore the relationship, actual or ideal, of poet and reader, the poem is the poem's subject – that is all.' The reader isn't distracted by contingent detail, what Davie calls 'the reek of the human'.[25] If Davie is right, it should be possible to use 'An Epistle to a Patron' as a remarkably pure example of how distinct and essential features of poetic composition relate to one another – according to Davie, in the absence of what Keats called any 'palpable design' on its readers. I want to examine Davie's proposition about the way the 'Epistle' works, to use a modified variant of it to produce an alternative description of the poem, and then to make some general points about the functions of intentional activity in non-narrative poetry on the basis of that alternative description.

To read a poem is to respond to the unfolding of a special kind of intentional act as a sequence of intentions in action. This is true when we read any work of literature, in any form or genre. But in the case of a narrative poem or play, much of the burden of conducting the writer's prior intention along the often devious channels of his intentions in action is born by the characters, in what I hope I have shown can be a startling variety of ways. But non-narrative poems can't be intentionally charged by characters for the simple reason that they don't contain characters in the sense of designing and willing agents who inaugurate actions and determine plots. In poems of this sort the author has to make direct and immediate use of the language that was mediately and indirectly used by the characters on the author's behalf in the others. Properties of that language, then, become the privileged and exclusive carriers of intentional energy in their own right.

Davie's account of 'An Epistle to a Patron' as an example of pure poetry opens with a quotation from Paul Valéry. Davie is discussing Mallarmé's theory of poetic syntax. I reproduce the Valéry passage, which will be familiar to some readers from its appearance in Elizabeth Sewell's *The Structure of Poetry*, if not from Davie's own book:

> In this – and I told him so one day – he approached the attitude of men who in algebra have examined the science of forms and the symbolical part of mathematics. This type of attention makes the structure of expressions more felt and more interesting than their significance and value. Properties of transformations are worthier than what they transform; and I sometimes wonder if a more general notion can exist than the notion of a 'proposition' or the consciousness of thinking no matter what.[26]

Syntax in poetry operates like mathematics when, as in Mallarmé's (and Valéry's) poetry and, Davie argues, in Prince's, 'its function is to please us in and for itself'. Advancing the view that there is no reason why the poem should ever stop, 'for it is plain that under the guise of architect speaking to patron the poet is speaking to his reader, and speaking about his poem even as he writes it', Davie strongly implies that the poem is *sui generis*, because the 'structures of expression' and the subject to be expressed turn out to be fundamentally the same. The act of writing the poem is the subject of its having been written.

I rely on Davie's commentary on 'An Epistle to a Patron' for three reasons. The first is that it is the only appreciation of the poem known to me and comes from an authoritative source. The second is that its content pretty well exactly matches the response I made to the poem when I first read it, and continue to make. The third is that it is a seriously incomplete account (after all, it is less than two pages long) in so far as it fails to explain some features of the poem that don't seem to bear out the main point Davie is making about its absoluteness and its purity. In fairness to Davie, then, I want to quote a part of the paragraph that immediately follows his account of 'An Epistle...' Then we shall have before us a satisfactory context for discussion of the poem in different terms from his, but, I think, leading to similar conclusions. Davie has said that 'An Epistle...' talks about itself because, in a poem in which 'the structures of expression are more interesting than the structures of experience behind them', that is the only alternative to being entirely meaningless:

> But this is true only so long as the poet is determined to make his poetry 'pure' and 'absolute'. For in Valéry's formula, it is not necessary that the structure of expressions should be the only source of interest in the poem, only that this should be more interesting than anything else. And even Valéry's formula is too narrow. For there is no reason why this sort of syntax... should be more than one source of pleasure among many others. It is poetic in that it gives poetic pleasure, and it differs from other kinds of syntax only in this – that the pleasure it gives has nothing to do with mimesis. On these terms, any amount of older poetry can be seen to employ syntax-like-mathematics and indeed this category becomes more crowded than any of the others.

Davie is commenting on syntax alone. I am trying to broaden the discussion in order to incorporate other features of poetic expression such as diction, rhythm and imagery. In these broader terms, what is being said is that most poems provoke responses from their readers both to what they are and to how they go about being what they are. 'What they are' is usually translated as 'subject', 'how they go about being what they are' as 'style'. But this leaves out the poem's intellectual, emotional and nervous impact on the reader. The presence of good poems like 'An Epistle...' within the corpus of English poetry, poems that according to Davie don't have a subject separate from their style, tends to bring out the inadequacies of conventional descriptions of poetic communication. If 'An Epistle...' makes the impact Davie says it does ('this splendid poem') then it has to be admitted that a strictly poetic effect of a not insignificant kind can be achieved without a 'fit' of style to subject, or of expression to attitude, being present at all.

A way of getting round this problem is to break down the monolithic character of words like 'subject' and 'style' so as to include within the definition of subject ('what a poem is about') the intention of the poet in creating the poem as expressed in the intentional act of his writing it. Valéry might have had something of the sort in mind when he emphasized what he called *'une abstraction motrice plus que philosophique'* as a characteristic feature of the *'lyrisme net et abstrait'* of the *'Cimetière marin'*,[27] or when he wrote of the same poem: *'je n'ai pas voulu dire, mais voulu faire, et . . . ce fut l'intention de faire qui a voulu ce que j'ai dit.'*[28] In her book on Valéry, Christine Crow translates the phrase *'une abstraction motrice'* as 'a motivating force',[29] which is precisely what I have in mind when I refer to a poem as an intentional act. The subject of Keats's 'Ode to a Nightingale' would commonly be described as 'the nightingale' or 'the nightingale's song' or 'what the nightingale's song means to Keats'. I am suggesting that the subject is also Keats's intention to write about each and all of these things as it is realized in the intentional act of writing the poem: the intentional act comprising both the intention to write about these things and the intention in writing that follows from it, but that might not in each and every respect meet the conditions of satisfaction of the prior intention.

The first of Keats's subjects I propose to designate the 'accusative' and the second the 'nominative' subject. The two adjectives have the effect of distinguishing between a subject thought of as being 'out there' – somehow separated from the operations involved in thinking about it – and a subject brought into the arena of a person's mental, emotional and nervous behaviour (which automatically involves its being drawn into the scope of his intentions). While the 'accusative' subject is generally accepted as being the subject of a poem, the 'nominative' subject is often confused with an aspect of a poem's style or with something that is just in the mind or consciousness of the poet and not really part of the poem at all. This is because the accusative subject is usually identified with all of the subject of a poem, and all aspects of intention are usually assumed to be private and subjective. Once these two misconceptions have been corrected, it is possible to look at how poems describe, or enact, their subjects in a more effective way.

Particularly it helps to explain how Prince's 'Epistle' works as a poem which in one sense effectively and conventionally 'fits' subject to style but which in another sense has no subject at all. I would describe it as a poem which opens with the announcement of the identity of its accusative subject and goes on to transform that accusative subject into a feature of style – to be precise, a continuous metaphor – which then acts as the means whereby the nominative subject is fully articulated throughout the poem. How, then, should we describe the accusative subject of 'An Epistle to a Patron'?

The poem is a dramatic monologue cast in the form of a letter written by an anonymous architect to a noble lord whom he hopes to persuade to become his patron. The letter form was used by Browning in monologues like 'Cleon' (the closest in tone and manner to Prince's poem) or 'Karshish' and by Pound in 'Exile's Letter' or 'The River-Merchant's Wife'. But the most cursory inspection will show how different the 'Epistle' is from any of these superficially comparable poems. Pound's monologues are free translations of the work of an eighth-century Chinese poet. They have something of the character of documents placed at a remove from the poet not only because they feign to have been written by an imaginary character, but also because in a sense they have been written by another real poet. Far from deploying an argument, these poems distil a mood – but a mood that is made over into an objective fact rather than poured out as a subjective expression. The reflective character of Pound's monologues, taken together with their sustained mood of pathos and nostalgia, marks them off sharply from anything we find in Prince.

Browning's poems are different too. In spite of the ebullience and spiritedness they share with Prince, there is no denying the difference that is made by the dramatic movement of the poems. Much of our pleasure in reading them springs from the way we build up a more and more complete picture of the situation from which the monologist speaks or writes. There is a specific time, place, problem, history, purpose and prompting event – and we discover what these are as the sequence of arguments, persuasions, descriptions and insinuations unfolds. Our discovery in its turn affects the way we respond to these arguments *etc.* The process of reading is exploratory and dynamic. There would be no difficulty in interpreting the intentional action of the poet here as it is displayed through the intentions in action of the characters he has invented to speak and write on his behalf. In Prince's poem, though, the writer has no name, and neither does his patron. We know by the middle of the sixth line what the speaker has it in mind to do, and that, in a straightforward and unambiguous sense, is what he does in the rest of the poem. There is no sequence of dramatic events unfolding during the remainder, and therefore no dramatic activity of discovery and interpretation on the reader's part as he moves from line to line and from sentence to sentence.

It might be more helpful to draw a comparison with one of Browning's spoken monologues, 'My Last Duchess', where the identity of the Duke and the Duchess is not disclosed and the tiny clue in the subtitle 'Ferrara' is far too enigmatic to provide the basis of a confident interpretation of the facts. Many candidates have been proposed for the Duke and Duchess, but the Oxford editors' conclusion that 'When the poem was finished Browning cannot have expected his readers to associate it with any particular Duke and Duchess' is almost certainly correct.[30] In 'My Last Duchess', though, the non-disclosure of the names doesn't have the effect of drawing attention away from the personalities and fortunes of the characters. On the contrary, the mystery of the speaker's identity and the anonymity of the 'Duchess' of whom he speaks set up in the poem's audience the same suspicions, uncertainties and a disposition to ask questions that they imagine are shared by the envoy to whom the Duke is speaking. There is a complicated story of love, jealousy, intrigue and murder teasingly hinted at but never fully disclosed in the Duke's negotiations about the dowry. Nothing so specific emerges from the architect's letter in 'An Epistle to a Patron'. There are no tantalizing gestures towards a half-hidden story, no hints of a plot lying behind the ostensibly straightforward address from artist to patron. How much detail, in fact, are we given about the writer and the recipient of Prince's letter – beyond the fact that the recipient is rich and powerful and the writer, in his own estimation, gifted and available? In other words, where are the equivalents of Neptune's sea horse and Fra Pandolph's hands, or, for that matter, the drawing aside of the curtain and the commands that 'all smiles stop together'?

Mention of these details from Browning's poem reminds us that his monologues are populated by many more characters than the speaker and the person spoken to. What would we make of the Bishop of St Praxed's without the presence in his mind of his old rival, Gandolf? Or Fra Lippo Lippi without the prior and his niece? Even in his most claustrophobic monologue, 'Andrea del Sarto', the situation between Andrea and Lucrezia would lose some of its dramatic tension and psychological complexity without the presence of the cousin and his whistle just outside the window. In

Prince's poem, however, there is no one but the writer and his patron. The only detail we are given about the writer's circumstances appears in lines 49–51, where he refers to his 'household consisting/Of a pregnant wife, one female and one boy child and an elder bastard/With other properties'. Here the last word, 'properties', tends to diminish the importance of the people it refers to, and in any case they contribute nothing by their presence to the poem. It could be argued that the disclosure of these facts about the writer actually subtracts from the reader's perception of his individuality rather than augments it. Turning to the patron, the only detail we are given about him is that he has under his protection an 'eastern hostage'. He too plays no dramatic role in the poem. Nothing changes in the writer's attitude to the patron, or what is assumed to be the patron's attitude to the writer, as a result of his casual appearance here. This merely two-way movement between writer and patron – with nothing between them in the nature of a shared understanding, a guarded secret, a past event in which they were both involved, a common ambition which they both pursue – has the effect not merely of shutting out the rest of the world, but of shutting out the patron too. His presence in the poem becomes not so much the object of the speaker's persuasion as the objectification of his desire. A character whose role never extends beyond the potential affording of opportunities for another is in danger of becoming identified with those opportunities. He is the metaphorical representation of the ambitions and desires of the speaker, a way for him to represent to himself what those ambitions and desires are when they are translated into a symbolic form. And since he is not represented as being in any accidental or contingent way separable or different from the ambitions and desires that drive him, he too disappears from the text as anything more than a tactical convenience, a way of arranging around a central point all those objects, images and activities to which he is made to refer.

The firm though optically enigmatic outlines of the Browning monologue have been dissolved by Prince in a way that makes both the writer and the recipient of the letter which is the poem's accusative subject disappear into the imagery, syntax and rhythmical deployment of its members. This could be seen as a development of the dramatic monologue in the modernist period during which Eliot, not Pound, is the representative figure. In early monologues like 'Prufrock' and 'Gerontion', Eliot removed the 'objective' presence of an action involving characters speaking and spoken to that had been essential to the successful working of the Browning monologue. This is very different from Prince, though the aim of both poets was what might appear superficially to be the same thing: the removal of character and action from a literary form that seemed to be held together by nothing else. The difference consists of what remained when they had done so. For Eliot, what remained was a rudderless 'floating' of mood and depleted passion; for Prince, a fresco of language that manages to survive the absence of both the wall on which it has been painted and the subject it is intended to represent.

This was the way I had read 'An Epistle to a Patron' before I encountered Davie's remarks on it, and I think nothing I have written runs against the current of what I take to be Davie's speculations about the poem. Now I want to throw a spanner into the works. Some readers will already have seen where the spanner has come from. It has come from a reading of Vasari's *Lives of the Artists*. Referring to Leonardo da Vinci's journey from Florence to Milan in 1482, the editor of Vasari's *Life* quotes from the letter of self-recommendation Leonardo wrote to Ludovico Sforza, Il Moro, which has come down to posterity in the *Codice Atlantico* in the Ambrosian Library. This is what Leonardo wrote in his letter to a very powerful and imposing patron:

> Most illustrious lord, having now fully studied the work of all those who claim to be masters and artificers of instruments of war... I will lay before you Lordship my secret inventions, and the offer to carry them into execution at you pleasure.
>
> An extremely light and strong bridge. An endless variety of battering rams. A method of demolishing fortresses built on a rock. A kind of bombard, which hurls showers of small stones and the smoke of which strikes terror into the enemy. A secret winding passage constructed without noise. Carved wagons, behind which whole armies can hide and advance...
>
> In time of peace, I believe myself able to vie successfully with any in the designing of public and private buildings, and in conducting water from one place to another. Item: I can carry out sculpture in marble, bronze, or clay, and also in painting I can do as well as any man be he who he may.

Again, I can undertake to work on the bronze horse, which will be a monument... to the eternal hour of the Prince your father, and the illustrious house of Sforza.[31]

It is impossible not to suppose that this real letter from Leonardo to Il Moro lies somewhere in the background of the anonymous epistle of Prince's poem. The opening address, the mixture of arrogance and subservience in the tone, the references to both war machines and architecture, the mention of 'secrets' in both opening paragraphs – all of these things point to the dependence of the fictitious on the real epistle. Add to them the way 'statutes/Admirable as music' reminds us that Leonardo travelled to Ludovico's court initially as a musician (he was sent by Lorenzo the Magnificent with the present of a silver lyre in the form of a horse's head), and the way the eloquent lines on real horses at lines 25–7 prompt reflections on Leonardo's equestrian monument to Francesco Sforza (Ludovico's father) which he modelled in clay in 1493. Also the hints on several occasions that the writer, like Leonardo, was rarely able to finish his projects: 'These flights with no end but failure,/ And failure not to end them' (lines 55–6), '. . . the panting mind/ Rather than barren will be prostitute' (78–9), *etc* And finally the unprovable but surely very strong feeling throughout the poem that it has a North Italian Renaissance setting. Take all these facts about the poem which tally either with the spirit and wording of the Milan letter itself or with circumstances we know attended its being written, and it is surely impossible to doubt the influence it exerted on the writing of the 'Epistle'. If we were to go further, into forbidden Livingston Lowes territory, we would also find that Prince is an expert on the Italian sources of English Renaissance poetry, has written a book on *The Italian Element in Milton's Verse*, and included in his second book of verse a long poem feigned to have been spoken by Michelangelo.

The evidence is, I think, conclusive. Prince depended quite heavily on the sense and the wording of the Milan letter when he wrote the poem. But this is not the same as to say that 'An Epistle to a Patron' is a fictitious letter supposed to have been written by Leonardo to Ludovico Sforza some time in 1482 and modelled on the real Milan letter in the *Codice Atlantico*. A poet can use a literary document with no prior intention of alerting his readers either to its influence on the writing of his own poem or even to its existence. The way literary references were used by Eliot in *The Waste Land* and by Pound in *The Cantos* might prompt us to suppose that we are always expected to detect their provenance because knowledge of their provenance significantly affects our response to the way they are used in the modern poem. But this doesn't have to be so. We probably overestimate the degree of familiarity with the detail of his Horatian original Pope expected of his readers. Not all of them were Swifts, Montagus and Burlingtons, and the appeal of many of Pope's translations was often one of what Felicity Rosslyn has called a 'generalised allusiveness'.[32] In any event the sort of work required of a modern reader apropos the Leonardo references in Prince's 'Epistle' would have been unthinkable to an eighteenth-century reader of Pope. But set aside for the moment the traditional precedents and concentrate on the comparison with Eliot. While it might have been reasonable of Eliot in the early 1920s to suppose that an educated readership would have a nodding acquaintance with Ovid, Dante and Shakespeare, would it have been equally reasonable of Prince in the early 1950s to suppose that an educated readership would be so familiar with the *writings* of Leonardo as to be able to pick up the reference to an obscure letter, seldom reproduced (though quoted in abridged form in Kenneth Clark's *Leonardo da Vinci*) and mainly useful as evidence of Leonardo's genius as an architect and engineer, not as a draughtsman and painter?

I confess I never knew the writer of 'An Epistle to a Patron' was Leonardo until I happened on the version of the letter in Clark's book. Before that, I had read the poem as an entirely fictitious letter from an anonymous Italian Renaissance architect and *universale uomo* to an equally anonymous wealthy patron. It was only after reading the letter in Clark that I began to pick out what one might call the Leonardesque elements in the poem. Prince deliberately leaves the identity of the actors unknown. Why not call the poem 'An Epistle from Leonardo to Il Moro' if he wanted us to think of it as such? He had no qualms 16 years later in calling his Michelangelo piece 'The Old Age of Michelangelo', not 'The Old Age of a Cantankerous Sculptor'. Then, why is there no reference to any one particular project Leonardo undertook while he was in Milan – if not the Horse, then perhaps 'The Last Supper'? And this puts us in mind of an even more spectacular omission. Even in the real letter, Leonardo mentions that 'also in painting I can do as well as any man'. Why is there not a single direct reference to Leonardo's painting in the poem? The lines about his love of light (28–41) suggest aspects of his painting. (Light 'clanged back by lakes and rocks' at line 34, for example, puts one in mind of 'The Virgin of the Rocks', 'The Virgin and Child with St Anne', even the 'Mona Lisa'.) On the positive side, why does the patron reserve an apartment for his 'eastern hostage' (what could this mean in relation to Il Moro's military and diplomatic activities?), and why is the writer burdened with a pregnant wife and the 'other properties' mentioned at line 50? (Surely Leonardo is unimaginable in the context of domestic liabilities described here?) Significantly, the only lines in the poem which seem to be offering some sort of detail about the actual lives of the writer and patron, as distinct from the life the writer aspires to live at the patron's court, emphatically point away from the identification of writer and patron with Leonardo and Ludovico Sforza. That is, if without benefit of Vasari, Clark and the *Codice Atlantico*, we ever imagined them to be so identified.[33]

Prince almost certainly used the Milan letter as a source for his own anonymous epistle. But the fact that he kept it anonymous, combined with other ambiguous features of the poem, suggests that he didn't want the scope and movement of the verse affected by the reader's noticing that this was so. Or it suggests that if the reader had noticed that it was so, Prince didn't want him to attribute to the source the sort of burden of significance he might quite rightly attribute to the lines about crowds flowing over London Bridge in *The Waste Land*, or men setting keel to breakers and going forth on the godly sea at the opening of the *Cantos*.

It is probably no more helpful to 'know' that Prince's model for his letter writer was Leonardo da Vinci than it is to know that, like his letter writer, Prince himself is fascinated by light and, also like him, he 'quite easily' sees himself as a 'successful active man,/ Architect, engineer or lawyer' (*Memoirs in Oxford*, III. ii, 2–6). The power of the 'Epistle...' comes from Prince's use of the Leonardo letter as a stimulus to composition, not just at the opening of the poem, but throughout. He uses its elevated tone and phraseology to launch the 'Epistle', and returns to it from time to time to keep the rest of the poem afloat and buoyant on a steady current of rhetoric. But the reader doesn't need to know that this is part of what he is doing. He senses the connection between letter and poem from the way the poem is articulated. No umbilical cord of irony, substantive implication or even cultural snobbery remains to provide what Prince – in the title of a later piece – was to call 'A Last Attachment'.

Indeed, the reference to the Leonardo letter could do more to blunt responses to the poem than to sharpen them. Features of the language which induce in the reader feelings of puzzlement, unease and subliminal satisfaction are too easily accounted for by reference to the 'source'. If Leonardo and Ludovico's silver lyre come too readily to mind after reading the phrase about 'statutes/ Admirable as music' at lines 2–3, then the half-concealed synaesthetic colouring of that phrase, brokenly supported by the way the 't' in 'statutes' is softened through its end-word connection with 'house' in the previous line, is lost. The wider dislocations of the second sentence (lines 6–12), in which military and civil arts are mysteriously combined ('civil structures of a war-like elegance' etc.), are too easily excused on the grounds that Leonardo, as artist and draughtsman, had drawn up plans for fortifications and offensive weapons before writing to Ludovico and refers to these in his letter. The superb lines on the horses in Ludovico's stables would carry less momentum if the abortive plan for a bronze equestrian statue got mixed up in one's appreciation of 'The copper thunder kept in the sulky flanks of your horse...' It seems likely to me that Prince inserted the line about the writer's household and the patron's hostage in order to deflect attention from the false Leonardan identification so that it would continue to be paid to other aspects of the poem that are more crucial to the effect it makes.

What is that effect? Briefly – the unfolding of an intention that has taken the form of a wish to persuade. This is expressed in terms that marry the persuasive force of the speaker's rhetoric with complementary characteristics of feeling and therefore of language. The characteristics of feeling might be described as: arrogance (or, better, *superbia* – perhaps justified, in a Renaissance setting, by the artistic accomplishments of the speaker); an awareness of the illimitable claims of the private person; dissatisfaction that the pressures of the world and the demands of the self are incompatible; self-abasement; glory in the material world and what can be made in it and of it. These and many others. But the point about them is that they are not displayed, as Browning would have displayed them, as identifiable features of the speaker's psychology. Instead, they are set at a distance from him. He sets them at a distance from himself. So they are used, not as ends in themselves, but as the emotional charges that give life to the words that express them. Consequently the aural, tonal and rhythmical properties of the word sequences are more musical than mimetic.

The most obvious aspect of the poem's metrical form is its long lines, normally of 15 to 19 syllables carrying from four to six stresses in a basically trochaic pattern. Elsewhere Prince has shown himself adept at taking over unusual metrical forms from other English poets and some Italian ones. *Memoirs in Oxford* is written in the five-line stanzas of Shelley's *Peter Bell the Third*; 'Afterword on Rupert Brook' adopts Bridges' hexameter syllabics from *The Testament of Beauty* and 'Poor Poll'; on other occasions he writes strambotti and many variants of sixteenth-century Italian canzone and seventeenth-century English Metaphysical stanzas; and he has used verse forms in the manner of Yeats, Browning and Tennyson. But 'An Epistle...' strikes me as entirely original, without the faintest shadow of previous use. It is brilliantly adapted to Prince's purpose in this poem, since it produces an energetically flowing rhythm that risks neither falling into prose cadences nor breaking apart into 'double measures' of octo- or decasyllables. Occasionally he inserts a shorter line, in the manner of Milton's half-lines in 'Lycidas' and the choruses of *Samson Agonistes*. These have the effect of creating a grave pause in the poem's movement, but they don't harshly interrupt the long-line rhythms because the stress pattern remains 'within bounds'. For instance line five – 'These few secrets which I shall make plain' – carries at least five full stresses, although it is only nine syllables long. It lacks the quantities of poetic feet carrying lighter, almost tripping rhythms, of the lines preceding and following it, and this establishes continuity with difference – which is what in their several ways most of the other lines are doing across their wider span.

The length of the line, combined with the trochaic metrical base and the absence of end-rhyme, creates many opportunities for half-rhyme and rhythmic echoes both inside individual lines and across lines which may be separated from each other by as many as three or four others. Take the passage on the horses, at lines 21–7:

... You may commission
Hospitals; huge granaries that will smile to bear your filial plunders,
And stables washed with a silver lime in whose middle tower seated
In the slight acridity you may watch
The copper thunder kept in the sulky flanks of your horse, a rolling
    field
Of necks glad to be groomed, the strong crupper, the edged hoof
And the long back, seductive and rebellious to saddles.

The rhythms of the verse are hammered out of the balances of mono- and polysyllabic words in each line. For example it is no accident that in the second line 'hospitals' and 'granaries' are both dactylic feet, separated by the heavily stressed 'huge', which carries forward the 'h' alliterative pattern over the join at the semicolon. Or take the last line. Both syllables in 'long back' are heavily stressed, requiring a slight pause between the words as well as a long pause at the comma after them. So – although there are only four syllables before the comma, and only two of them are stressed – because the stresses come where they do, they give the impression of aurally lengthening the syntactic member. After the comma-pause, the phrase about saddles seems to have the same value, in spite of its extra word, and in spite of the fact that three of the five words are polysyllabic (two of them trisyllabic). This is because of the more urgent forward-moving rhythm created by the combination of dactylic and trochaic feet, and the way each word rushes to meet the next in order to complete the sound pattern that links 'seductive' to 'saddles' by way of the middle-stress syllable of 'rebellious'. This syllable opens and closes on the same consonants that introduce each of the stressed syllables in the phrase before the comma. Stronger sound-links between the more widely separated members of the sentence connect 'filial' in the first line with 'silver' and 'middle' in the second; and 'middle' is half heard again in 'saddles' – a near rhyme – at the end of the last line. 'Plunders' half-rhymes with 'thunder', and these echoes are reinforced by 'copper' before 'thunder', and 'crupper' in the next line. Throughout the six lines there are less pronounced but still effective alliterative links in the succession of words with sibilant openings – a discreet sound-device that is brought out into the open only in the last half-line. Elsewhere the same care is taken to link different semantic areas of the verse through correspondences that are rarely mimetic. In the first sentence, for example, the musical parallels between 'opulence', 'promises', 'parasites' and 'prospering' have only weak correspondences in the links, or lack of them, between the meanings of these words.

In most circumstances readers will agree with Pope's dictum that the sound must seem an echo to the sense. But the sound and sense in the passage under consideration seem to have little to do with each other. Back to Valéry and Mallarmé: 'This type of attention makes the structure of expressions more felt and more interesting than their significance or value.' And this is true of the syntax as well as of the metrical patterning: the rhythmic units are played off against, within and between very carefully demarcated sentences, clauses and phrases. The balance of declarative indicative clauses and subordinate, often participial phrases affects the way we hear repetitions of similar sounds as we move from sentence to sentence along the grammatical trajectory of the poem.

But there is some semantic point in the very disjunction of sound and sense in the 'Epistle'. It brings to poetic life the significance of what lies within the covert references to Leonardo's correspondence with Ludovico Sforza, while ensuring that it remains covert. And this is how connections between separate members of a mechanical project which remain rudimentarily utilitarian when applied prosaically to building canals or measuring piers between window frames are transformed into a self-sustaining aesthetic or poetic beauty when incorporated within the shapes and patterns created by the organization of sound in verse. Syntax and rhythm don't work alone; diction and imagery also play a part. In lines 6–12, for example, 'plotted' controls the way we respond to the activity of warfare in the next line ('ordinary engines/ Of defence and offence' *etc*) and of building in the line after ('building/ . . . . some are courts of serene stone' *etc*), though in the first sense the meaning of the word is more abstract (= 'calculating') than in the second (= 'making a design', 'working out how to use a plot of ground'). The sentence mixes civil and military, beautiful and functional, in unpredictable ways: 'courts of serene stone' and 'citadels of brick' producing a confused mêlée of the artistic and the practical. This conjunction of apparently discordant words and images, producing on an intellectual level the same sort of effect that synaesthesia produces on a sensory one, beckons towards an aesthetic order that exists independently of the raw materials out of which it has been formed. It is the verbal reflection of what the writer of the epistle wants to achieve in plastic terms for his patron. Occasionally, though, the words he uses to display that ambition simultaneously reflect it and refer to it in the verse – as in phrases like 'hanging together/Like an argument' at lines 14–15, or 'your closet waiting not/Less suitably shadowed than the heart' at 19–20. These are only the most explicit comparisons in a poem which elsewhere makes use of works like 'orator', 'lines', 'figured' and 'the habits of numbers' as inset metaphors within the encompassing metaphor of the architect writing his epistle.

The activities described in Prince's 'Epistle' – engineering, building, inventing weapons of war – are end-directed. In Kant's terms they might accidentally succeed in displaying a dependent beauty. The passage from intention of doing to purposes completed is a direct one, with none of those creative deflections from the straight course of practical efficiency produced by a too close involvement in certain kinds of intention in doing. The conditions of satisfaction are met by the careful observance of conventional practices. The illocutionary force of the activity is entirely commensurate – no more and no less – with the perlocutionary effect of its completed performance. As an expression of an intention realized in an intentional action that is the writing of the poem, Prince's 'Epistle' is end-directed merely in the sense that it finally comes to a point of rest. But it does this only after extending itself far beyond anything that was required by the announcement of its purpose in the first six lines of the poem. All that was required of it by those lines, all that they entailed, was a list of the author's accomplishments, together with a suitable tone of deference to the addressee. But it does far more than this. As an illocutionary speech act it carries so many and such various combinations of forces, that the move from illocutionary intention to perlocutionary effect is suspended and delayed over a long period of time. During that time the language works inefficiently from the point of view of realizing its declared purpose, but effectively from the point of view of making the most of the intentions in action created out of its dilatoriness in reaching a single conclusion.

This is because of the mainly instrumental status of the accusative subject – which is the architect's letter to the patron, shorn of the character it actually displays in the poem but complete in its function of impressing upon the patron the practical skills of its author. However, the way the epistle elaborates on those practical skills persuades the reader to interpret what is written in the poem as a metaphorical expression of something other than an architect's letter, even an architect's letter invented by a real author for fictional purposes such as revealing the character of the architect. That metaphorical expression has something to do with the way the poet's intention to write his poem is realized in the sequence of intentions in action, represented by everything that exists in the poem. This is the poem's nominative subject: the containment of the accusative subject of the poem – the letter conceived of as a sort of précis – in the intentional activity Prince brings to bear on it.

The letter never existed as a creation of Prince's imagination, even if he knew of something very like it in Leonardo's letter to Ludovico Sforza, which was indeed efficient and end-directed. The nominative subject is everything we have in the poem. The accusative subject is merely what it would have been if a poem hadn't been made out of it. It would have been a useful document, with a clear intention aimed at the achievement of a specific and limited purpose. Like Kant's tulips or Copleston's rose, it becomes supremely purposive, but without purpose. To complete itself was not above all things what it sought to do. But at last it completes itself all the same. The precise nature of the final appeal, then – to patron or to reader – depends on whether our eye is fixed upon the accusative or the nominative subject. Both poem and epistle submit themselves to judgment. But how differently we interpret these last lines, and how differently we interpret that crucial word 'judgement' in them in our double capacity as readers of the architect's letter and readers of Prince's poem:

> For my pride puts all in doubt and at present I have no patience,
> I have simply hope, and I submit me
> To your judgement which will be just.

Epilogue

> A sense of direction is not the same thing as a goal-orientated system.
> *Ornithologist on the navigation of homing pigeons*, BBC Radio News, 1 July 1988.

Almost without exception, recent literary theory has been anti-essentialist, mocking what, in his *Preface to Paradise Lost*, C. S. Lewis called the doctrine of the unchanging human heart. It remains a fact that some important things about being human do seem to be essential. To get at them, however, we have to put to one side impermanent belief systems like Virgilian imperialism and Elizabethan codes of honour (Lewis's examples), or even, perhaps, what appear to be the less temporally circumscribed emotions – like the will to power and romantic love – of which these belief systems have been the periodic 'carriers'. What is permanent about the behaviour of human beings, distinguishing them from all other forms of life and producing the essential subject of their art and literature, doesn't offer itself quite so explicitly as a subject. It has more of the appearance of a formal category. Nonetheless it *is* a subject, in the sense that it is what all writing is *about*, as well as what in formal terms it *is*. I think that this essential property can be identified as whatever is produced when an impulse to what Coleridge called 'the shaping spirit' and a self-reflexive sense of purpose join together in the act of creating a poem, play or novel.

Writing and our response to writing – the solicitation of one consciousness by another – is the type of our activity, in so far as we must act purposefully if we are to act as human beings at all. Particular purposes wax and wane in importance over time. For Michael Tippett, the meaning of the fall of Troy and the lesson of the anger of Achilles bear only a passing resemblance to what they were to the Greeks of Homer's time. In *Omeros*, Derek Walcott transforms Homer's epic characters into the fishermen of his native St Lucia, and their values and goals in life change accordingly. The long quest of Don Quixote for an ideal self-image is shrunk by Kafka into a single paragraph attributing the whole enterprise to a whim of Sancho Panza.[1] Marthe Robert has shown how the exigencies of the twentieth century transformed Cervantes' comic epic into the parodic narrative of Kafka's *Castle*.[2] In his role as a chivalric knight, K finds that 'his good will serves no purpose, for there is no longer any object called the Grail or anything else that represents a goal or even a direction for all men of the same period and culture.' More recently he has resurfaced as a political refugee (in David Wheldon's *The Viaduct*) and a concert pianist (in Kazuo Ishiguro's *The Unconsoled*). In both cases his particular purposes have changed as much as a sense of purposiveness in his actions, however baffled or lacking an appropriate object, has remained constant.

The particular purposes served by Homer's, Cervantes' and Kafka's accounts of their subject are only the refracted images of a larger sense of the purposiveness of life, and of individual lives. Reading, viewing, responding to a narrative is unimaginable without a sense of purpose lying behind the words, persuading us to ask questions about why this happens, what all this is *for*. Most readers would agree, I think, that what a text is *for* is, finally, inseparable from what it *is* – a literary form imbued with intentional activity and wrought out of words that are at one and the same time patterns of sound, Saussurean sign-concepts and instruments of reference. But is the essence of the text only what Shelley called a 'feeble shadow of the original conception' and Hardy an 'irradiated conception' incapable of being expressed through the clumsy figures of narrative?[3]

UNCLE ISAK'S STORY

We are half way through the fifth act of Ingmar Bergman's film *Fanny and Alexander* (1982).[4] 'Uncle' Isak has spirited the children away from Bishop Vergerus's gloomy palace and brought them to his mysterious house of masks and puppets. This is the centre of the film's world of the creative imagination. In it, Isak tells a story which in some ways sounds like a hasidic tale, in others like a parable by Borges or Kafka. A young man is travelling along an endless road across a stony plain. He is tired, thirsty, alone in the midst of crowds of people travelling alongside him. He is oblivious to the sounds of flowing water and the flash of the streams and pools between which he threads his way. He meets an old man who tells him about the water. It issues from springs high in the mountains, springs that have been formed out of the collective hopes, fears, longings and prayers of mankind. The crowds that press forward on the road foolishly ignore these springs, though everyone has at some time heard of them. 'Most people remain anxiously on the road in the glaring light.' Why this is so, the old man cannot say. 'Perhaps they believe they will reach their unknown destination by evening.' 'Which unknown destination?', the young man asks. 'Probably there is no destination,' the old man replies. 'It is deception, or imagination.' Earlier the destination had been described as 'the goal of their pilgrimage', and the young man's 'final goal' which he has forgotten. It is better to retrace one's steps to the mountains and the springs, though it is not easy. 'Next morning the youth sets out with the old man to seek the mountain, the cloud, the forest and the rippling springs.'

The parable throws some light on the matters that have been discussed in this book. The man who tells it, Uncle Isak, is a good man, the most effectively good man in the film. Some time in the past he had been the lover of Mrs Ekdahl, Fanny's and Alexander's grandmother. By the time Fanny and Alexander have been rescued from the bishop's palace, however, their father has been dead for a year, and it is easy for us to assimilate the dead father we have known to the dead grandfather we have never seen. As a result, Mrs Ekdahl (Helena) comes to stand for a sort of mother to the children and Isak a sort of father. This impression is reinforced by the fact that the actors who play the parts of Isak and Oscar (the father) – both long-established members of Bergman's team – are much the same age. Isak, then, becomes one of three father figures to Fanny and Alexander. There is the real father, Oscar; the step-father, Bishop Vergerus; and the proxy father, Isak. However, there is one respect in which Isak does not share in the parental status of the other two men, and this exclusion has a significant effect on our response to his tale.

Fanny and Alexander is a story about relationships between sons and fathers (one son and three fathers). It is also a story about *Hamlet*. The Ekdahl family own a theatre and produce plays in it. Near the beginning of the film they are shown rehearsing *Hamlet*. The real father, Oscar, plays Hamlet's father's ghost. Later he dies; his widow, Emilie, remarries; and the new husband becomes a wicked stepfather to her children. Emilie warns Alexander against thinking of Bishop Vergerus as if he were Claudius in the play. But Alexander ignores her warning and we can understand why. Oscar twice reappears to Alexander, and once to Emilie as a ghost. The film, therefore, both explicitly and implicitly, challenges its audience to draw parallels between the Hamlet story and the Ekdahl story. Where the bishop is concerned, these are clear and unambiguous. He is a Claudius figure. But Isak is not so straightforward. His relationship with Mrs Ekdahl identifies his role with that of Old Hamlet, though Helena is not actually Alexander's mother and Oscar is the ghost. Also, Isak's relation to Emilie is pointedly neither that of an Old Hamlet nor of a Claudius to Gertrude. Bergman goes out of his way to keep Isak and Emilie apart (they scarcely ever meet in the film), and her youthfulness is contrasted with his age even more than it was with Oscar's. The fact that he is a Jew also sets him to one side of the family relationships.

Isak's role in the film is to provide a sort of shadowing around the more brightly illuminated outlines of the other male figures. He exists as an alternative to each of them – the dead grandfather, the recently deceased father, the all too living stepfather – until the occasion of the fire at the bishop's palace for which, in the alternative, magical narrative, he seems to bear some responsibility. The ambiguity of the role he plays influences our response to the actions he performs. The magical device by which he hides the children in an old wooden chest, seconds before we see them locked into the room overhead, is the only one Bergman can find to move the story forward at that time. The more comically realistic methods of the real uncles, Carl and Gustav, are by contrast ineffective. So in order to be freed from the obstructions of its own plot, the narrative has to be interrupted by Isak's magic. At each re-viewing this remains a puzzling feature of the film. Isak's recommendation, through the lips of the old man in his parable, of a return to those mysterious springs and sources is equally puzzling and ambiguous. The endless way is like the plot of the film, its characters moving forward towards uncertain goals and unknown destinations. The hidden springs, diverting the narrative to a source of spiritual replenishment that exists apart from its own forward movement, is like the magic displacement that puts Fanny and Alexander in two different places at the same time (out of time).

The roads we travel in our response to drama, film and fiction have the same kind of double direction. The regression to mysterious sources of illumination and refreshment is countered by a movement in the opposite direction, a progress along a road that stands for no single aim and leads to no certain destination. The old man's advice to the young man, Uncle Isak's rescue of the children from their imprisonment in the bishop's palace – something like this lies at the back of much post-Romantic commentary on art, and much contemporary art as well. The reader's attention travels along both the horizontal and vertical axes of a text – responsive both to the metaphoric possibilities of roads not taken, words not used, and to the metonymic actualities of a destination approached, words and images following one another in an orderly sequence. But to conceive of words as entirely metaphorical, unrestricted in their range of imaginative reference to anything more than the intentional foreclosures of consciousness, is to settle in the end either for incommunicable mysteries or for messages that are communicable only by virtue of their transparency and their self-cancellation.

Uncle Isak's conjuring trick and the old man's retirement to the mountains and the springs are effective only in association with the forward movement of the rest of the film, the travelling onward of all those other people. If we were to view the story of Fanny and Alexander only from the perspective of Uncle Isak and the old man, it would soon begin to dissolve under our gaze. We need to see it from Alexander's, and Bergman's, own point of vantage. This encompasses the magic of the return to origins, but it isn't controlled by them. The realization of the artistic impulse depends on Bergman's intentional activity spread out along the evolving narrative of the whole film. The modifications of the *Hamlet* story are part of it. So are the activities of the other father figures. So are the contrasted images of the theatre, the cathedral and the Ekdahl household. Among these images, Uncle Isak's magic toyshop and the mysterious parable he reads to Alexander inside it have their place. Their place is to try to expand into the rest of the film, and to fail to do so. Ultimately, they are absorbed into the metonymic flow of the rest of the narrative. This, Isak says, 'is deception, or imagination' – just the words we might have used, unprompted, to describe the magical episodes it contains.

In the end, though, these are acceptable, if not fully comprehensible, as phases in the intentional activity of the whole film. Apart from their abruptness and mystery, they have no privileged status in the narrative. *All* of it is deception, or imagination. The conjunction 'or' is carefully selected to register the ambiguity of phrase. Are they the same, or are they alternatives? Is purposiveness in human affairs fully to be accounted a constitutive or a regulatory principle? That is what art cannot tell us. Perhaps it is a matter on which the theory of art should be silent too.

Notes

CHAPTER 1 GOOD INTENTIONS

1. P. F. Strawson, *The Bounds of Sense*, London, 1966. 0.273. See also Thomas Nagel, 'What is it like to be a bat?', in *Mortal Questions*, Cambridge, 1979, pp. 165–80. For a thorough discussion of human thinking about animals, see Mary Midgeley, *Beast and Man: The Roots of Human Nature*, London, 1980.
2. Stuart Hampshire, *Thought and Action*, London, 1959, p. 57. See also Michael Leahy, *Against Liberation*, London, 1991. For a different view of intention in animals, see James Rachel, *The End of Life*, Oxford, 1987, and Peter Singer, 'Taking life: animals', chapter 7 of *Practical Ethics*, Cambridge, 1997.
3. All references to *The Critique of Judgement* (*CJ*) in this chapter and in Chapter 3 are to the translation by J. H. Bernard, London, 1892.
4. Paul Valéry, *Cahiers*, Paris, CNRS, 1957–61, XVIII p. 350.
5. Ted Hughes, '*Orghast*: Talking Without Words', *Vogue*, December 1971, pp. 95–7.
6. 'For a Birthday', *Fighting Terms*, London, 1954. Other poems by Gunn are 'On the Move', 'Human Condition' and 'Vox Humana', from *The Sense of Movement*, London, 1957; 'Lights Among Redwood', *My Sad Captains*, London, 1961; 'Touch' and 'Back to Life', *Touch*, London, 1967.
7. Italo Calvino, *If on a winter's night a traveller*, London, 1981, p. 71.
8. John Holloway, *Narrative and Structure: Exploratory Essays*, Cambridge, 1979, pp. 111–12. Compare Stuart Hampshire on 'co-construction' in 'A composer's world', *Modern Writers and Other Essays*, London, 1969, pp. 174–83.
9. *CJ*, I, 14, p. 59. At I, 7, p. 47 Kant refers to different people's responses to the colour violet, and to the tone of wind instruments compared with that of strings. For a more detailed account of this aspect of Kant's aesthetic, see below pp. 156–7.
10. Wittgenstein's move from the picture theory of language to the anti-theory of language games and forms of life reawakened personal doubts and uncertainties that preceded composition of the *Tractatus*. See *Notebooks 1914–1916*, eds G. H. von Wright and G. E. M. Anscombe, Oxford, 1961, especially entries under 4.2.16 at p. 30: 'Isn't the thinking subject in the last resort mere superstition? . . . Where in the world is a metaphysical subject to be found?'
11. Preface to Jacques Derrida, *Of Grammatology*, Baltimore, Md., 1976. The argument for Derrida's abolition of the essential subject in a

philosophy-of-language context can best be studied in his essay 'Signature event context' and contested in John Searle's reply, in vol. 1 of *Glyph*, 1977, pp. 177–97 and 198–208. Derrida had the last word in a reply to Searle in vol. 2 of *Glyph*, 1977, pp. 162–254. The most recent of several much-needed demolition jobs on Derrida's casuistry in this and other statements of his position is ch. 3 ('The fallacies of hermeticism') of Wendell V. Harris, *Literary Meaning: Reclaiming the Study of Literature*, London, 1996, pp. 50–64.

12. Freud has been pressed into the service of an intentional theory of narrative in Peter Brooks' *Reading for the Plot: Design and Intention in Narrative*, Oxford, 1984. See for example, p. xiii on 'the dynamic aspect of narrative . . . seeking in the unfolding of the narrative a line of intention and a portent of design . . . .', and pp. 286–7: 'plots must generate force: the force that makes the connection of incidents powerful, that shapes the confused material of a life into an intentional structure . . .'

13. The clearest exposition of Lacan's use of these terms is in Ellie Ragland-Sullivan, *Jacques Lacan and the Philosophy of Psychoanalysis*, London and Canberra, 1986, pp. 42–62.

14. James M. Mellard, *Using Lacan, Reading Fiction*, Urbana, Ill. and Chicago, 1991, p. 56.

15. Raymond Tallis, *Not Saussure: A Critique of Post-Saussurean Literary Theory*, London, 1988, 1995, p. 157. For a more recent exposé of Lacan's abuse of scientific method and terminology, see Alan Sokal and Jean Bricmont, *Intellectual Impostures: Postmodern Philosophers' Abuse of Science*, London, 1998.

16. For some literary implications and applications of Lacan's transformation of Freud's *Fort! Da!* experiment, see Christine van Boheemen, *The Novel as Family Romance: Language, Gender, and Authority from Fielding to Joyce*, Ithaca, NY, 1987, esp. p. 18: 'The utterance originally intended as substitute thus becomes its own goal and satisfaction.'

17. Michael Payne, *Reading Theory: An Introduction to Lacan, Derrida, Kristeva*, Oxford, 1993, pp. 116–17.

18. Two of the most powerful recent arguments against deconstruction are Sean Burke, *The Death and Return of the Author*, Edinburgh, 1992, and Raymond Tallis, *Not Saussure*, (see n.15 above).

19. Ragland-Sullivan, op. cit. p. 14.

20. For value and signification in Saussure, see John Holloway, 'Language, realism, subjectivity, objectivity', in Lawrence Lerner (ed.), *Reconstructing Literature*, Oxford, 1983, pp. 60–80; and for sense and reference in Saussure, see Tallis, 'Statements, facts and the correspondence theory of truth' in *Not Saussure*, op. cit., pp. 235–50.

21. Perhaps the contemporary philosopher whose theoretical preoccupations are closest to Gunn's is Richard Rorty. His 'transcendental hermeneutics' shares with Gunn's poetry a suspicion of foundational approaches to personhood and of essential descriptions of consciousness. Kant is singled out for special critical consideration in his

Philosophy and the Mirror of Nature, Princeton, NJ and Oxford, 1980. See esp. pp. 137, 148–9 and the whole of 'Part Three'.
22. Geoffrey Hill, 'Poetry as "menace" and "atonement"', *The Lords of Limit*, London, 1984, p. 7.
23. John Fletcher, *Claude Simon and Fiction Now*, London, 1975, p. 88.
24. Roland Barthes, 'The death of the author', *Image–Music–Text*, trans. and ed. Stephen Heath, London, 1977, pp. 210–11.
25. M. H. Abrams, *Doing Things with Texts: Essays in Criticism and Critical Theory*, New York and London, 1989, pp. 275–6.
26. Sean Burke, op. cit., p. 148.
27. Ibid., pp. 14–15.
28. Tallis, *Not Saussure*, p. 205.
29. Italo Calvino, 'A sign in space', *Cosmicomics*, London, 1993, p. 32.
30. Edward Said, *Beginnings*, 2nd edn, New York, 1985, pp. 372 and 319–20.
31. Stuart Hampshire, op. cit., p. 135.
32. Christopher Ricks, *Essays in Appreciation*, Oxford, 1996, p. 51. Cf. Raymond Tallis, *Not Saussure*, op. cit., p. 234: 'There is a darkness at the heart of intention and there is an inescapable indeterminacy in their relation even to the actions that seem most precisely to realize them.'
33. W. K. Wimsatt and M. C. Beardsley, 'The intentional fallacy', *The Sewanee Review*, vol. LIV, Summer, 1946, pp. 468–88; reprinted in *The Verbal Icon*, Lexington, Ky., 1954, pp. 3–18.
34. Ibid., p. 3.
35. Paul de Man, *Blindness and Insight: Essays in the Rhetoric of Contemporary Criticism*, 2nd edn, London, 1983, p. 27.
36. What follows contradicts the argument, advanced most cogently by Richard Wollheim in *Art and its Objects*, London, 1980, that little is gained in aesthetic theory by pointing to the different senses in which the word 'intention' is used. Wollheim argues that the differences between what he calls 'ulterior' and 'immediate' intentions lie only on the surface of the word: 'only the shortest distance below the surface, these different "senses" of the same word are inter-related' (73). But see the arguments of Anscombe and Searle at pp. 73 ff. and 85 ff. below.
37. Nigel Williams, *The Wimbledon Poisoner*, London, 1990, p. 298.
38. Eric Griffiths, *The Printed Voice of Victorian Poetry*, Oxford, 1989, pp. 52–4.
39. Allan Rodway, *The Truths of Fiction*, London, 1970, pp. 123 and 113.
40. Ludwig Wittgenstein, *Culture and Value*, Oxford, 1970, pp. 58e–59e.
41. 'The Frontiers of Criticism', in *On Poetry and Poets*, London, 1957. But compare the difficulties arising out of a related passage from the same essay at p. 166 below.
42. *Philosophical Investigations*, 3rd edn, Oxford, 1968, p. 165e.
43. Robert Audi, *Practical Reasoning*, London and New York, 1989, p. 127; Donald F. Gustafson, *Intention and Agency*, Dordrecht, 1986, pp. 8 and 203.
44. E. D. Hirsch, *Validity in Interpretation*, New Haven, Conn., 1967; P. D. Juhl, *Interpretation: An Essay in the Philosophy of Literary Criticism*,

Princeton, NJ, 1980; Stanley Fish, *Is There a Text in This Class?: The Authority of Interpretive Communities*, Cambridge, Mass., 1980.
45. Steven Knapp and Walter Ben Michaels, 'Against theory', *Critical Inquiry*, Summer 1982, vol. 8, no. 4; reprinted in W. J. T. Mitchell (ed.), *Against Theory: Literary Studies and the New Pragmatism*, Chicago and London, 1985.
46. William C. Dowling, 'Intentionless meaning', in ibid., pp. 89–94.
47. A. D. Nuttall, *The Stoic in Love*, Hemel Hempstead, 1989, pp. 197–8. See also John Kemp, 'The work of art and the artist's intention', *British Journal of Aesthetics* vol. 4, 1964, pp. 146–54.
48. Sir John Salmond, *Jurisprudence*, 8th edn, London, 1930, p. 393.
49. C. K. Ogden and I. A. Richards, *The Meaning of Meaning*, 10th edn, London, 1949, p. 20.
50. Ragland-Sullivan, op. cit., p. 59.
51. *The Concept of Mind*, London, 1949; Harmondsworth, 1963 p. 26.
52. For Ryle on motive and intention see J. K. Jenkins, 'Motive and intention', *Philosophical Quarterly*, vol. 15, 1965, pp. 155–64.
53. In B. Williams and A. Montefiore, *British Analytic Philosophy*, London, 1966, p. 215
54. See, for example, N. S. Sutherland, 'Motives as explanations', *Mind*, vol. LXVIII, April 1959, pp. 145–59; and L. W. Beck, 'Conscious and unconscious motives', *Mind*, vol. LXXV, April 1966, pp. 155–79.
55. See Max Black (ed.), *Philosophy in America*, London, 1965, pp. 221–39; reprinted in Searle's (ed.) own *The Philosophy of Language*, London, 1971.
56. Ibid., p. 46.
57. J. L. Austin, *How to Do Things with Words*, Oxford, 1962.
58. P. F. Strawson, 'Intention and Convention in Speech Acts', in Searle, op. cit., p. 31.
59. Searle, op. cit., p. 8. See also 'Meaning, communication and representation', in R. E. Grandy and R. Warner (eds), *Philosophical Grounds of Rationality: Intentions, Categories, Ends*, Oxford, 1986, on 'primary-meaning intention' and 'communication intention'.
60. In Searle, op. cit., p. 31.
61. Perhaps best expressed as a state in which an action is performed. See Jenkins fn 52 above, and Grice, 'Intending is not a process or activity, but rather, perhaps, a state', in 'Intention and uncertainty', *Proceedings of the British Academy*, vol. LVII, 1971.
62. Ohmann appears to contradict this point of view. He refers to 'the suspension of normal illocutionary force' in a literary work, but the emphasis on 'normal', and the admission of 'mimetic illocutionary force' suggests that the contradiction might be more apparent than real. See Richard Ohmann, 'Speech acts and the definition of literature', *Philosophy and Rhetoric*, vol. 4, 1971, pp. 1–19.
63. Cf. Grice's reservations about the applicability of conversational maxims to non-declarative speech acts, in H. Paul Grice, *Logic and Conversation*, excerpts in P. Cole and J. L. Morgan (eds), *Syntax and Semantics Vol. 3: Speech Acts*, New York, 1975. See also Patrick Suppes,

'The primacy of utterer's meaning', in Grandy and Warner, op. cit., pp. 109–29, and Grice's own essays on 'utterer's meaning' in *Studies in the Way of Words*, Cambridge, Mass., 1989.

64. This distinction conforms with arguments of recent speech-act theorists and linguists that there are no significant formal differences between 'natural' and 'literary' narrative. See William Labov, 'Narrative analysis: oral versions of personal experience', *Essays on the Verbal and Visual Arts: Proceedings of the 1966 Annual Spring Meeting of the American Ethnological Society*, Seattle, 1967, pp. 12–45; and Mary Louise Pratt, *Towards a Speech Act Theory of Literary Discourse*, Bloomington, Ind., 1977.

65. *Biographia Literaria*, chs xiv and xv.

66. See Pratt, op. cit., p. 141: 'there is clearly a level of analysis at which utterances with a single point or purpose must be treated as single speech acts . . . or "texts".' Also Charles Fillmore, 'Pragmatics and the description of discourse', in Fillmore, G. Lakoff and R. T. Lakoff (eds), *Berkeley Studies in Syntax and Semantics*, 1974, V, pp. 1–21 (quoted in Pratt, op. cit., p. 142).

67. Alvin I. Goldman, *A Theory of Human Action*, Englewood Cliffs NJ, 1970. For alternative philosophical approaches to the relation between motive and action, see R. S. Peters, *The Concept of Motivation*, London, 1958, and Donald Davidson, 'Actions, reasons, and causes', *The Journal of Philosophy*, vol. LX, 1963, and 'The logical form of action sentences', in N. Rescher (ed.), *The Logic of Decision and Action*, Pittsburg, 1967. More recent attempts to tackle the problem are listed on p. 211 of Paisley Livingston, *Literature and Rationality: Ideas of agency in theory and fiction*, Cambridge, 1991. There are very few instances of the application of action theory to literary hermeneutics. Apart from Livingston himself, see Charles Altieri, *Act and Quality: A Theory of Literary Meaning and Humanistic Understanding*, Amherst, Mass., 1981; Uri Margolin, 'The doer and the deed: action as a basis for characterization in narrative', *Poetics Today*, vol. 7, 1986, pp. 205–25; and Bijoy H. Boruah, *Fiction and Emotion: A Study in Aesthetics and the Philosophy of Mind*, Oxford, 1988. Altieri's book includes commentary on Goldman (in ch. 2) and applies action description to an interpretation of a poem by William Carlos Williams (in ch. 4).

68. Kenny, *Action, Emotion and Will*, London, 1963, pp. 171–86.

69. Hampshire, *Thought and Action*, pp. 100–1. For the application to literature of Hampshire's view of intention here, see the 'Introduction' to his *Modern Writers, and Other Essays*, op. cit., London, 1969, pp. xiii–xiv.

70. Goldman, op. cit., pp. 16–17.

71. *Ibid.*, p. 19.

72. *Ibid.*, p. 59.

73. For detailed explanations of the four methods of level-representation, see ibid., pp. 22–30.

74. '. . . man is only a recent invention, a figure not yet two centuries old, a new wrinkle in our knowledge, and . . . he will disappear as soon as that knowledge has discovered a new form.' Michel Foucault, 'Preface'

240 Notes

to *The Order of Things: An Archaeology of the Human Sciences* [1966], trans. A. M. Sheridan Smith, London, 1970.
75. See *Culture and Value*, op. cit., p. 57e.

CHAPTER 2 SHAKESPEARE

1. G. K. Hunter (ed.), *Macbeth*, London, 1967, p. 40.
2. Marvin Rosenberg, *The Masks of Macbeth*, Los Angeles and London, 1978, p. 263.
3. The editors of the original (1912) and the 1951 Arden editions of the play. George Steevens and Sir Herbert Grierson are the main proponents of the alternative reading.
4. Barry Unsworth, *Morality Play*, London, 1995, p. 55.
5. Ludwig Wittgenstein, *Philosophical Investigations*, p. 165e.
6. Though see Part I, para. 337. Also Wittgenstein *Brown Book*, para. 147. Both passages quoted in ch. 1 of John Casey, *The Language of Criticism*, London, 1966, p. 9. Also Wittgenstein, *Zettel*, 2nd edn, Oxford, 1981, pages 582–600, on voluntary and intentional movement.
7. John Casey, op. cit.
8. R. S. Peters, *The Concept of Motivation*, London, 1958, p. 37.
9. Carl Ginet, *On Action*, Cambridge, 1990, p. 76.
10. *Cf.* Madeleine Doran's syntactic approach to the same issue in 'Iago's "if": an essay on the syntax of *Othello*', in E. M. Blistein (ed.), *The Drama of the Renaissance: Essays for Leicester Bradner*, Providence, RI, 1970, pp. 69–99.
11. G. E. M. Anscombe, *Intention*, Oxford, 1957, pp. 37–47.
12. An alternative paradigm of intentional action is Ginet's exposition of 'nested groups of actions in which one action is the core of a more complex action, which in turn is the core of a still more complex action, and so on.' The 'simple basic action', which must be a mental action (in the case of Anscombe's parable, the man moving his hand up and down in a deliberate effort to operate the pump), 'is intentional by its very nature'. The intentionality of the events that follow on the performance of the simple basic action varies from case to case, but I take it that all four events described in my slimmed-down version of Anscombe's parable are intentional. The last two events in the example from Searle (see p. 82 above), however, are not. 'The possibility of failing to be intentional . . . comes in only with complex actions, only when an action contains a layer of consequence or circumstance that the agent could conceivably fail to expect or fail to be aware of when performing the core action.' See Ginet, op. cit. pp. 73–4 ff.
13. Iago has three more soliloquies after the arrival in Cyprus: the first at II. iii. 326–52; the second at III. iii. 318–26; and the third at V. i. 11–22. The last two of these arise out of particular exigencies of situation (the finding of the handkerchief, the arrangement of the fight between Cassio and Roderigo), and are correspondingly specific in what they say. The first and longest of them adds most to this commentary on

the earlier soliloquies – especially its last three lines, in which Iago's plans are expressed in terms that are as extreme as their meaning is obscure.
14. An alternative and complementary account of Iago's behaviour appears in A.D. Nuttall, *A New Mimesis: Shakespeare and the Representation of Reality*, London, 1983, pp. 131–43. In Nuttall's reading of the play, 'Iago is not just motivated, like other people. He *decides* to be motivated. He concedes that he has no idea whether Othello has had sexual relations with his wife [Emilia]. He simply opts, in a vacuum, for that as a possible motive.'
15. For an alternative (though tantalizingly brief) version of the relation between motive, intention and purpose in Iago's behaviour, see Wendell V. Harris, *Interpretive Acts: In Search of Meaning*, Oxford, 1988, pp 47–8. Harris speculates on the subject using the terminology of intention, purpose and project found in Charles Altieri's *Act and Quality: A Theory of Literary Meaning and Humanistic Understanding*, Amherst, Mass., 1981.
16. Donald F. Gustafson, *Intention and Agency*, Dordrecht, 1986, p. 106. Another of Gustafson's concepts we might use in analysis of Iago's behaviour is what he calls the 'truth- gap': 'Intentions are propositions with a difference; they contain a special type of predicate or special category of predicate. These special predicates have the feature that they are satisfiable by future contingent states-of-affairs only. Consequently propositions with ineliminable predicates of this category exhibit "truth-gaps"; they are neither true nor false until the worlds to which they are temporarily indexed are parts of the actual world' (p. 112).
17. See T. M. Raysor (ed.), *Samuel Taylor Coleridge: Shakespearean Criticism*, vol. I, London, 1960, p. 44.
18. The terminological link between intention and purpose, made possible by discriminating between 'coherence' and 'planning' models of intention, has been argued by J.A. Passmore, 'Intentions', *Aristotelian Society Supplement*, 1955. Casey (op. cit., p. 149) agrees that 'when a critic says that a writer "means" or "intends" something . . . he is talking in terms of . . . the "coherence" . . . model' which 'seeks only to pick out a purposive pattern in behaviour'.
19. Barbara Everett accurately describes Macbeth's activity in this scene when she writes that 'His journey towards murder is both unerring and pathless', in *Young Hamlet: Essays on Shakespeare's Tragedies*, Oxford, 1989, p. 82.
20. *Intentionality: an Essay in the Philosophy of Mind*, Cambridge, 1983 – See especially ch. 3, 'Intention and Action'. For a critical account of the literary implications of Searle's theory, see Eckhard Lobsien, 'The "intentional fallacy" revisited: on John Searle's *Intentionality*', *Comparative Criticism: A Yearbook*, vol. 7, Cambridge, 1985, pp. 279–88.
21. J. P. Stern discusses the relation between a writer's 'meaning' and his character's 'purpose' in similar terms in *On Realism*, London, 1973, p. 81: 'Whatever the end in view, and whether or not it is clearly conceived before it is achieved, it is bound to shape the fiction . . . in

22. something like the same way that the details of [what happens in it] are shaped by *its* end; though in each case not the end only but it *and* the available means do the shaping.'
22. See *Tolstoy on Shakespeare*, trans. V. Tchertkoff, New York, 1906.
23. B. M. Eichenbaum, *Skvoz' literatu* (*Through Literature*), Leningrad, 1924, p. 81.
24. The passages from L.S. Vygotsky are taken from *The Psychology of Art* (*Psikologiya Isskusstva*) trans. MIT Press, Cambridge, Mass. and London, 1971.
25. John Dover Wilson, *What Happens in Hamlet*, 3rd edn, Cambridge, 1951, p. 52. See also Nigel Alexander, *Poison, Play and Duel*, London, 1971, pp. 32–3: 'the nature of the Ghost is intended to be an open question.'
26. H. de Vocht (ed.), *Jasper Heywood and His Translations of Seneca's Troas, Thyestes and Hercules Furens*, Louvain, 1913, p. 40.
27. For the bearing of Elizabethan views on both ghosts and revenge on the interpretation of *Hamlet*, see Eleanor Prosser, *Hamlet and Revenge*, Stanford, Calif., 1967, and Peter Mercer, *Hamlet and the Acting of Revenge*, London, 1987, pp. 121–69.
28. Richard Levin, *New Readings vs Old Plays*, Chicago and London, 1979, pp. 153–6.
29. Vygotsky, op. cit., p. 192.
30. Gustafson relates the notion of *things under a description* to differences between two ways of individuating action: the 'Multiplier' view, which holds that distinct action properties distinguish different actions; and the 'Unifier' account, which holds that the same action can exemplify different and distinct action properties or action descriptions. See op. cit., p. 16.
31. In *The Legacy of Wittgenstein*, Oxford, 1984, p. 146
32. More recently, Gilbert Harman has supplied examples of circumstances showing that 'someone can do something intentionally without intending to do it'. See 'Willing and Intending', in R. E. Grandy and R. Warner (eds), *Philosophical Grounds of Rationality: Intentions, Categories, Ends*, Oxford, 1986, pp. 363–80.
33. Raymond Tallis, *The Explicit Animal: A Defence of Human Consciousness*, London, 1991, p. 213. Elsewhere Tallis reminds us that it can work the other way round, and that we can lose as well as gain a sense of intention in the processes of realizing it: 'our intending selves may temporarily absent themselves from what we are doing. We do not continuously intend even our most deliberate actions throughout their duration' (*Not Saussure: A Critique of Post-Saussurean Literary Theory*, London, 1988, 1995, p. 217).
34. For the reader's response to this process of composition of a text, see Frank Kermode, 'Divination', in *An Appetite for Poetry*, London, 1989, pp. 152–71
35. See Arthur Sherbo (ed.), 'Johnson on Shakespeare', Yale edn of *The Works of Samuel Johnson*, vol. VIII, pp. 980–1. Modern critics, by contrast, insist that the speech is connected both in Hamlet's mind and on his tongue. See especially Harry Levin, *The Question of Hamlet*, New

York, 1959, and Stephen Booth, 'On the value of *Hamlet*', in Norman Rabkin (ed.), *Reinterpretations of Elizabethan Drama*, New York, 1969, pp. 171 and 175.
36. See, for example, J. Middleton Murry in *Things to Come: Essays*, London, 1938, p. 231: 'What is "to be or not to be" is not Hamlet, but Hamlet's attempt upon the king's life.' Quoted in Harold Jenkins' Arden edn of *Hamlet*, London, 1982, p. 486.
37. The phrase order in T.H. Visser's commentary on the verb 'to be' reinforces the claim: 'Where *to be* is used as an independent, notional verb its principal meanings are: to occupy a place, to exist; to happen, to come to be; to remain or go as in an existing condition.' In *An Historical Syntax of the English Language*, Part One, Leiden, 1970, p. 160.
38. Stuart Hampshire, *Thought and Action*, London, 1959, p. 167.

CHAPTER 3 COLERIDGE AND KANT

1. Quotations from *Wilhelm Meister* are from Thomas Carlyle's translation of 1824, Book V, chapters 5 and 7
2. See Barbara Everett, *Young Hamlet: Essays on Shakespeare's Tragedies*, Oxford, 1989, p. 28.
3. These two phrases roughly correspond with Kenny's 'motive' and 'intention' in ch. 4 of *Action, Emotion and Will*, London, 1963, especially pp. 86–9. For a different approach see L. W. Beck in 'Conscious and unconscious motives', *Mind* vol. LXXV, April 1959, on 'motive answers' and 'intention answers' to questions about reasons for actions.
4. J. K. Jenkins' 'Motive and Intention', *Philosophical Quarterly* vol. 15, 1965, pp. 155–164, examines Anscombe's argument in the light of Kenny's distinction between backward-looking and forward-looking reasons, in *Action, Emotion and Will*, op. cit., pp. 90–2.
5. Alvin I. Goldman, *A Theory of Human Action*, Englewood Cliffs, NJ, 1970, p. 52.
6. W. K. Wimsatt and M. C. Beardsley *The Verbal Icon*, Lexington, Ky., 1954, pp. 11–12.
7. On the history of Idealist transformations of Kant's aesthetic, see Andrew Bowie, *Aesthetics and Subjectivity: from Kant to Nietzsche*, Manchester and New York, 1990. For the link between Kant and Heidegger, see Hubert Dreyfus (ed.), 'Introduction' to *Husserl, Intentionality and Cognitive Science*, Boston, Mass., 1982, pp. 19–27.
8. See A. M. Quinton, 'Absolute idealism', in A. Kenny (ed.), *Rationalism, Empiricism, and Idealism*, Oxford, 1986, pp. 124–50.
9. J. Shawcross (ed.), *S. T. Coleridge, Biographia Literaria* (hereafter *BL*), vol. 1, London, 1907, p. 99.
10. E. L. Griggs (ed.), *The Letters of Samuel Taylor Coleridge* (hereafter *CL*), vol. I, Oxford, 1956, p. 209.
11. *Ibid.*, p. 284.
12. *Ibid.*, vol. II, pp. 677–703.
13. *Ibid.*, p. 706.

14. Rosemary Ashton, *The German Idea: Four English Writers and the Reception of German Thought 1800–1860*, Cambridge, 1980, pp. 42–3.
15. H. J. Jackson and George Whalley (eds), *Samuel Taylor Coleridge, 'Marginalia'*, vol. III, London, 1992, p. 241.
16. *CL*, vol. III, Oxford, 1959, p. 35.
17. R. Wellek, *Kant in England*, Princeton, NJ, 1931, pp. 109–11. A more recent commentary on Kant's influence on Coleridge's thought is Paul Hamilton, *Coleridge's Poetics*, Oxford, 1983, chapters 2 and 3.
18. Ashton, op. cit., pp. 49–50.
19. R. A. Foakes (ed.), *Samuel Taylor Coleridge: Lectures on Literature 1808–1819*, vol. 1, pp. 30 and 84.
20. T. M. Raysor (ed.), *Samuel Taylor Coleridge: Shakespearean Criticism*, vol. 1, p. 181n.
21. Foakes, op. cit., p. 115.
22. Wellek, op. cit., p. 110.
23. Raysor, op. cit., vol. 2, p. 201n.
24. Alice D. Snyder, *Coleridge on Logic and Learning*, New Haven, Conn., 1929.
25. *CL*, vol. III, p. 360.
26. 'Essay On the Principles of Genial Criticism', 1814, repr. in *Samuel Taylor Coleridge, Shorter Works and Fragments*, eds. H. J. Jackson and J. R. de J. Jackson, Princeton, NJ, 1995, p. 365.
27. Norman Fruman, *Coleridge, the Damaged Archangel*, London, 1972, p. 171.
28. James Engell, 'Coleridge and German Idealism: first postulations, final causes', in Richard Gravil and Molly Lefebure (eds), *The Coleridge Connection: Essays for Thomas McFarland*, London, 1990, pp. 165–6.
29. *BL*, p. 100
30. Rosemary Ashton, *The Life of Samuel Taylor Coleridge: A Critical Biography*, Oxford, 1996, p. 250.
31. See especially his descriptions of Determinate and Indeterminate Situations in the *Philosophy of Fine Art*, vol. 1, trans. F. P. B. Osmaston, London, 1920, pp. 266–72, and his commentary on the unity of dramatic action, *ibid.*, vol. 4, pp. 263ff. For Hegel on poetry and drama, see chapters ii and viii of J. Kaminsky, *Hegel on Art*, New York, 1962.
32. For Kant on the sublime, see Paul Crowther, *The Kantian Sublime: from Morality to Art*, Oxford Philosophical Monographs, Oxford, 1989. Recent treatments of the subject include Rudolf Makkreel, 'Imagination and temporality in Kant's Theory of the Sublime', *Journal of Aesthetics and Art Criticism*, vol. 42, no. 4, 1984, and Patricia M. Matthews, 'Kant's Sublime: a form of pure aesthetic judgement', *Journal of Aesthetics and Art Criticism*, vol. 54, no. 2, 1996. For a deconstructive reading, see J. F. Lyotard, *Lessons on the Analytic of the Sublime*, trans. Elizabeth Rottenburg, Stanford, Calif., 1994.
33. *CL*, vol. VI, p. 895.
34. Mary Warnock, *Imagination*, London, 1976, p. 49.
35. M. H. Abrams, *Doing Things With Texts: Essays in Criticism and Critical Theory*, New York and London, 1989, p. 166. He is quoting K. P. Moritz, *Schriften zur Aesthetik und Poetik*, ed. Hans Joachim Schrimpf, Tübingen, 1982.

36. Instead, see Crowther, op. cit., pp. 41–77.
37. In Mary McCloskey, *Kant's Aesthetic*, London, 1987, p. 94.
38. In Stephan Körner, *Kant*, Harmondsworth, 1955, pp. 180–2, 200–7.
39. For a critical account of the role of schema in Kant's theory of Imagination, see A. R. White, *The Language of Imagination*, Oxford, 1990, p. 45 (and the rest of ch. 7). Denis Crowley has drawn my attention to foreshadowings of an autonomy theory of the Imagination in changes he made between the first and second editions of *CPR*. See also Sarah Gibbon, *Kant's Theory of the Imagination*, Oxford, 1994.
40. *CJ*, I, 6, p. 45 and I, 22, p. 77.
41. *Ibid.*, I, 6, p. 45.
42. *Ibid.*, I, 17, p. 73
43. Richard Wollheim, *Art and its Objects*, London, 1980, p. 112.
44. *CJ*, I, 40, p. 138.
45. Roger Scruton, *Kant*, Oxford, 1982, p. 85.
46. McCloskey, op. cit., p. 105.
47. For Nelson Goodman, see *Languages of Art: An Approach to the Theory of Symbols*, Oxford, 1969.
48. *CJ*, I, 45, p. 149.
49. See, for example, D. W. Crawford, *Kant's Aesthetic Theory*, Madison, Wis., 1974, p. 134: 'Thus, somewhat paradoxically, natural beauty pleases us ultimately because it is like art – it seems designed for our contemplation of it.'
50. Frederick Copleston SJ, *A History of Philosophy, vol. vi: Wolff to Kant*, London, 1968, p. 360.
51. See, for example, *CJ*, I, 58, pp. 193 and 195.
52. Quoted by John Holloway in *Narrative and Structure: Exploratory Essays*, Cambridge, 1979, p. 115.
53. Ludwig Wittgenstein, *Culture and Value*, Oxford, 1970, p. 56e.
54. Mary Warnock provides a clear account of the bearing of aesthetic ideas on ideas of reason in *Imagination*, London, 1976, pp. 61–70. See also Eva Schaper, 'Taste, sublimity and genius: the aesthetics of nature and art', in Paul Guyer (ed.), *The Cambridge Companion to Kant*, Cambridge, 1992, pp. 367–93.
55. Scruton, op. cit., p. 88.
56. Warnock, op. cit., p. 42: 'In the *CPR* the ideas of reason are introduced in an almost entirely negative way. They stand for what we *cannot* conceptualise, and thus for what we *cannot* in any way experience in the world.'
57. For Kant on aesthetic ideas, see especially *CJ*, I, 1, 49. But this needs to be read in conjunction with his remarks on the intellectual sublime at I, 1, 29, pp. 111–12.
58. Crawford, op. cit., pp. 93–4. The passage quoted is from *CJ*, II, 1, 64, pp. 216–17.
59. Cf. Sean Burke on Derrida and the author as the 'transcendental signified': 'The text is read as natural theologians read nature for marks of design, signs of purpose' (*The Death and Return of the Author*, Edinburgh, 1992, p. 23).
60. See *CJ*, I, 16, p. 66.

61. L. S. Vygotsky, *Thought and Language*, trans. E. Hanfmann and G. Vikar, Cambridge, Mass., 1962, p. 146.
62. Recent developments in discourse analysis frequently emphasize similarities between ordinary conversation and the use of language in literature: see Mary Louise Pratt, *Towards a Speech Act Theory of Literary Discourse*, Bloomington, Ind., 1977. For the inappropriateness of the application of notions of 'aesthetic response' to both modes of discourse, see A. L. Becker, 'Beyond translation: estheics and language description', in Heidi Byrnes (ed.), *Contemporary Perceptions of Language*, Washington DC, 1982, pp. 124–8, and Deborah Tannen, *Talking Voices: Repetition, Dialogue and Imagery in Conversational Discourse*, Cambridge, 1989.
63. Donald Davidson's definitions of what he calls 'prior theory' and 'passing theory' and the relations between them seem to me especially appropriate to descriptions of these sorts of conversation. See 'A nice derangement of epitaphs', in R. E. Grandy and R. Warner (eds), *Philosophical Grounds of Rationality: Intentions, Categories, Ends*, Oxford, 1986, pp. 157–74 (especially pp. 167–9).
64. See, for example, *CJ*, I, 1, 48, p. 154: 'But if the object is given as a product of art and as such is to be declared beautiful, then, because art always supposes a purpose in the cause (and its causality), there must be at bottom in the first instance a concept of what the thing is to be.'
65. The link between abstract art and Romantic organicism is clear in the pronouncements of both European and American expressionist artists. For an early example, see Wassily Kandinsky on 'the spiritual accord of the organic with the abstract element' in *Concerning the Spiritual in Art* (1911), English translation, New York, 1977, p. 31.
66. Wittgenstein, *Culture and Value*, op. cit., p. 79e.
67. T. S. Eliot, *The Use of Poetry and the Use of Criticism*, London, 1933, p. 87.
68. The image of a bow and arrow, and the trajectory of the arrow, is a familiar one in the philosophy of intentional action: 'A human being's action is essentially constituted of means towards an end; it is a bringing about of some result with a view to some result. "With a view to", or "in order to", are unavoidable idioms in giving the sense of the notion of an action, the *arrow* of agency passing through then present and pointing forward in time; [For behaviour to be counted as intentional human action] there is . . . a requirement of connectedness, of a *trajectory* of intention that fits a sequence of behaviour into an intelligible whole, intelligible as having a direction, the direction of means towards an end', Hampshire, op. cit., pp. 73 and 146, my italics).
69. T. S. Eliot, 'The frontiers of criticism', in *On Poetry and Poets*, London, 1957, p. 110.

CHAPTER 4 MILTON, STERNE, PRINCE

1. For details of the theological background, see R. T. Kendall, *Calvin and English Calvinism to 1649*, Oxford, 1979. For the application to *Paradise*

Lost, see John Stachniewski, *The Persecutory Imagination: English Puritanism and the Literature of Religious Despair*, Oxford, 1991, ch. 8, pp. 332–84: on God's forbidding Satan's 'retirement from purposive action' (p. 367) and Satan's 'active choice having to be exercised in accordance . . . with a sense of destined purpose' (p. 376).

2. E. M. W. Tillyard, *Milton*, London, 1930; Stanley Fish, *Surprised by Sin: the Reader in Paradise Lost*, London and New York, 1967.

3. Another way of explaining Milton's intentions here is through correspondence between the representation of God in the epic narrative and the representation of God in the Bible. This is in turn explained through the principle of accommodation, according to which the representation of God in the Bible shows us how he wishes to be understood by mankind. Milton argues this in *De Doctrina Christiana*. See *Complete Prose Works* VI, ed. D. M. Wolfe, New Haven, Conn. and London, 1953–82, pp. 133–5, and Barbara Lewalski, *Paradise Lost and the Rhetoric of Literary Forms*, Princeton, NJ, 1985, pp. 111–14.

4. Michael Lieb, *The Dialectics of Creation: Patterns of Birth and Regeneration in Paradise Lost*, Cambridge, Mass., 1970, p. 67.

5. Arnold Stein, *Answerable Style*, Minneapolis, 1953, p. 127.

6. Lieb, op. cit., p. 101.

7. Leland Ryken, *The Apocalyptic Vision of Paradise Lost*, Ithaca, NY and London, 1970, pp. 168–9.

8. Hence the accuracy of John Peter's insistence on God's lack of effort in Book III: 'the figure is impressive, glancing from Earth to Hell with effortless circumspection; his foresight reaching out no less effortlessly into the dimension of Time' (*A Critique of Paradise Lost*, London, 1970, p. 11).

9. Dennis Burden, *The Logical Epic: A Study of the Argument of Paradise Lost*, Cambridge Mass., London, 1965, p. 25.

10. For commentary on the link between sexual and narrative activity in *Tristram Shandy*, see Dennis Allen, 'Sexuality/textuality in *Tristram Shandy*', *Studies in English Literature*, vol. 25, no. 3, 1985, pp. 651–70, and Roy C. Caldwell, '*Tristram Shandy*, bachelor machine', *The Eighteenth Century: Theory and Interpretation*, vol. 34, no. 2, 1993, pp. 103–14.

11. A naive relation between Sterne and Tristram as authors of the text formed the basis of most criticism of *Tristram Shandy* from Johnson to Traugott (John Traugott, *Tristram Shandy's World: Sterne's Philosophical Rhetoric*, Berkeley, Calif., 1954). It was questioned by Wayne C. Booth in several essays emphasizing Sterne's manipulation of the self-conscious narrator. See especially 'Did Sterne complete *Tristram Shandy*?', *Modern Philology*, vol. 48, 1950–1, pp. 172–83, and 'The self-conscious narrator in comic fiction before *Tristram Shandy*', *PMLA*, vol. 67, 1952, pp. 163–85. Booth's approach, supplemented by such critics as Holtz (1970), Swearingen (1977) and Lamb (1989), has exercised a profound and inhibitive influence on recent criticism of *Tristram Shandy* – as represented, for example, in Melvyn New's (ed.) 'New casebook', London, 1992 (which reprints Booth's 1952 essay).

12. D. W. Jefferson's 'Tristram Shandy and the tradition of learned wit', *Essays in Criticism*, vol. 1, 1951, pp. 225–48, opened up a rich seam of

fictional and non-fictional source material which almost all later critics of Sterne have made use of and many have continued to mine.

13. Trim's story has attracted very little critical attention, though the reading of Yorick's sermon has been commented on at length, e.g. in Andrew Wright, 'The artifice of failure in *Tristram Shandy*', *Novel*, vol. 2, 1968–9, Brown University, Providence, RI, Spring, 1969, and J. P. Hunter, 'Shandean interruptions', *Novel*, 4, 1970–1, Winter 1971, pp. 132–46. Max Byrd draws attention to sausage making as a risqué joke in which 'language replaces sexual action in *Tristram Shandy* ('Unwin Critical Library' edition), London, 1985, p. 134. There is a suggestive paragraph on the Jew's widow in Booth's 'Did Sterne complete *Tristram Shandy*?', op. cit.

14. Cf. Epistemon's story of the woman of Smyrna, in *Gargantua and Pantagruel*, Book III, ch. 44, or Sindbad the sailor's account of his misfortunes to Sindbad the porter in the story of Sindbad from the *Thousand and One Nights*. Rabelais' influence on Sterne has often been noted, but I have found no references to the *Arabian Nights*, though these stories had been available in an anonymous English translation from 1708, and in French in the translation of Antoine Galand from 1717.

15. The Russian formalist distinction between *fabula* and *suzet* might be applied to Sterne's recording of this incident. Disappointingly, Victor Shklovsky's essay on *Tristram Shandy* has little to say about Trim's stories, including the one about his brother Tom and the Jew's widow. See L. T. Lemon and M. J. Reis (eds), *Russian Formalist Criticism*, Lincoln, Nebr., 1965.

16. Poe's essay is the subject of an interesting argument about contemporary literary theory in Patrick Parrinder, *The Failure of Theory: Essays on Criticism and Contemporary Fiction*, Brighton, 1987, pp. 5–17.

17. For a detailed account of this aspect of the novel, see Peter Briggs, 'Locke's *Essay* and the tentativeness of *Tristram Shandy*', *Studies in Philology*, vol. 82, 1985, pp. 494–517.

18. Cf. Gabriel Josipovici on the coincidence of verbal and material analogies in *Tristram Shandy*, II. 7, in *Writing and the Body*, London, 1982, pp. 117–18.

19. Samuel Beckett, *Molloy, Malone Dies, The Unnamable*, London, 1959, pp. 301–2.

20. *Confessions*, Book XII. 15 (Pusey's translation, in the 'Everyman' edn, pp. 287–8).

21. *Ibid.*, XI. 6 and 7 (pp. 256–7).

22. Two essays, both in Melvyn New's (ed.) 'New casebook', op. cit., reinforce this view at the level of verbal detail. See Sigurd Burckhardt, '*Tristram Shandy*'s Law of Gravity', *ELH*, vol. 28, 1961, pp. 70–88 – on Sterne's exposure of the naivety of eighteenth-century faith in Locke's doctrine of simple ideas expressed in simple words; and Donald R. Wehrs, 'Sterne, Cervantes, Montaigne: fideist skepticism and the rhetoric of desire', *Comparative Literary Studies*, vol. 25, 1988, pp. 127–51 – on Sterne's deferral of narrative fulfilment through the invention of 'unpredictable events and unregulated linguistic connotations'.

Notes

23. F. T. Prince, *Collected Poems*, Manchester, 1993.
24. Donald Davie, *Articulate Energy*, London, 1955, pp. 92–3.
25. *Ibid.*, pp. 161–5.
26. Paul Valéry, *Variété*, vol. III, Paris, 1948, p. 28. See Elizabeth Sewell, *The Structure of Poetry*, London, 1951, pp. 151–2.
27. André Gide – Paul Valéry, *Correspondance 1890–1942*, in Paul Valéry, *Oeuvres*, vol. 2, ed. J. Hytier, Paris, 1977 [1960], 1687.
28. 'Au sujet du Cimetière Marin', *Oeuvres*, vol. 1, 1975 [1957], 1503.
29. Christine M. Crow, *Paul Valéry and the Poetry of Voice*, Cambridge, 1982, p. 201.
30. Ian Jack and Rowena Fowler (eds), *The Poetical Works of Robert Browning*, vol. III, Oxford, 1988, p. 185.
31. Ludwig Goldscheider, *Leonardo da Vinci: Paintings and Drawings*, London and New York, 1954, p. 17. The translation here is as in Clark, *Leonardo da Vinci*, pp. 45–6. Vasari claims that the Duke invited Leonardo to Milan, not that Leonardo invited himself – as the letter shows.
32. Felicity Rosslyn, 'Heroic couplet translation: a unique solution?', *PN Review*, vol. 23, no. 1, Sept.–Oct. 1996, p. 16. On the use of personal reference in Pope's poetry see Howard Erskine-Hill, *The Social Milieu of Alexander Pope*, New Haven, Conn., 1975.
33. In July 1993, the Palazzo Pubblico in Siena mounted an exhibition of the work of Francesco di Giorgio, an elder contemporary of Leonardo, whose projects for Federico Montefeltro in Urbino are a mirror image of those described in Leonardo's letter to Ludovico Sforza. There is no reason, outside our assumption of Prince's knowledge of this letter, for us to suppose that Leonardo is any more certainly the model for the poem's supplicant than is Francesco. Either or none of them will do to fill out the identity of the writer.

EPILOGUE

1. Franz Kafka, 'The truth about Sancho Panza', *Description of a Struggle and Other Stories*, Harmondsworth, 1979, p. 112.
2. Marthe Robert, *The Old and the New*, trans. Carol Cosman, Berkeley, Los Angeles and London, 1977, p. 214.
3. For Shelley see *The Defence of Poetry*; for Hardy see his letter to Coventry Patmore, 11 November 1886, collected *Letters of Thomas Hardy*, vol. I, eds R. L. Purdy and M. Millgate, Oxford, 1978, p. 157.
4. *Fanny and Alexander*, trans. Alan Blair, New York, 1983; Harmondsworth, 1989.

Select Bibliography

Full references to all books discussed are contained in the footnotes. The list below is of the principal primary and critical texts I have used, and the ones most frequently referred to in the course of the argument.

LITERARY REFERENCES

Ingmar Bergman, *Fanny and Alexander* (trans. Alan Blair), Harmondsworth, 1989.
Thom Gunn, *Fighting Terms*, London, 1954.
———, *The Sense of Movement*, London, 1957.
———, *My Sad Captains*, London, 1961.
———, *Touch*, London, 1967.
John Milton, *The Poems of John Milton* (eds John Carey and Alastair Fowler), London, 1968.
F. T. Prince, *Collected Poems*, Manchester, 1993.
William Shakespeare, *Complete Works* (ed. Peter Alexander), London and Glasgow, 1951.
———, *Othello* (ed. M. R. Ridley), London, 1958.
———, *Macbeth* (ed. G. K. Hunter), Harmondsworth, 1967.
———, *Othello* (ed. Kenneth Muir), Harmondsworth, 1968.
———, *Hamlet* (ed. Harold Jenkins), London, 1982.
Lawrence Sterne, *The Life and Opinions of Tristram Shandy* (ed. Christopher Ricks), Harmondsworth, 1967.

PHILOSOPHICAL REFERENCES

G. E. M. Anscombe, *Intention*, Oxford, 1957.
J. L. Austin, *How to Do Things with Words*, Oxford, 1962.
A. L. Goldman, *A Theory of Human Action*, Englewood Cliffs, NJ, 1970.
H. P. Grice, *Studies in the Way of Words*, Cambridge, Mass. 1989.
Stuart Hampshire, *Thought and Action*, London, 1959.
Immanuel Kant, *Critique of Pure Reason* (trans. and ed. J. M. D. Meiklejohn), London, 1876.
———, *Critique of Practical Reason and other works on the theory of Ethics* (ed. T. K. Abbott), London, 1879.
———, *Critique of Judgement* (ed. J. H. Bernard), New York, 1951.
———, *The Critique of Judgement* (ed. J. C. Meredith), Oxford, 1952.

Anthony Kenny, *Action, Emotion and Will*, London, 1963.
——, *The Legacy of Wittgenstein*, Oxford, 1984.
C. K. Ogden and I. A. Richards, *The Meaning of Meaning* (10th edn) London, 1949.
John Searle (ed.), *The Philosophy of Language*, London, 1971.
John Searle, *Intentionality: an Essay in the Philosophy of Mind*, Cambridge, 1983.
P. F. Strawson, *The Bounds of Sense*, London, 1966.
L. S. Vygotsky, *The Psychology of Art* (trans. MIT Press), Cambridge, Mass. and London, 1971.
Mary Warnock, *Imagination*, London, 1976.
Ludwig Wittgenstein, *Tractatus Logico Philosophicus* (trans. C. K. Ogden and F. P. Ramsay), London, 1922; (trans. D. F. Pears and B. F. McGuinness), London and New York, 1961.
——, *Notebooks, 1914–16* (eds G. H. von Wright and G. E. M. Anscombe, trans. G. E. M. Anscombe), Oxford, 1961.
——, *Philosophical Investigations* (3rd edn), Oxford, 1968.
——, *Culture and Value* (ed. G. H. von Wright in collaboration with N. Heikki), Oxford, 1980.

LITERARY CRITICAL REFERENCES

Sean Burke, *The Death and Return of the Author*, Edinburgh, 1992.
John Casey, *The Language of Criticism*, London, 1966.
Samuel Taylor Coleridge, *Biographia Literaria* (ed. J. Shawcross), London, 1907.
——, *Collected Letters*, vols I–IV (ed. E. L. Griggs), Oxford, 1956–9.
——, *Shakespearean Criticism* (ed. T. M. Raysor), London, 1960.
——, *Lectures on Literature 1808–1819* (ed. R. A. Foakes), London, 1971.
——, *Shorter Works and Fragments* (eds H. J. Jackson and J. R. de J. Jackson), Princeton, NJ, 1995.
Donald Davie, *Articulate Energy: An Enquiry into the Syntax of English Poetry*, London, 1955.
T. S. Eliot, *The Use of Poetry and the Use of Criticism*, London, 1933.
——, *On Poetry and Poets*, London, 1957.
Eric Griffiths, *The Printed Voice of Victorian Poetry*, Oxford, 1989.
Wendell V. Harris, *Literary Meaning: Reclaiming the Study of Literature*, London, 1996.
John Holloway, *Narrative and Structure: Exploratory Essays*, London, 1979.
W. J. T. Mitchell (ed.), *Against Theory: Literary Studies and the New Pragmatism*, Chicago and London, 1985.
A. D. Nuttall, *A Stoic in Love*, Hemel Hempstead, 1989.
Raymond Tallis, *Not Saussure: A Critique of Post-Saussurean Literary Theory*, London, 1988, 1995.
Paul Valéry, *Cahiers*, Paris, 1957–61.
W. K. Wimsatt and M. C. Beardsley, *The Verbal Icon*, Lexington, KY, 1954.

Index

act trees, 46–8
action theory, recent developments in, 239 n
activity words, *see under* Kenny, Anthony
act-properties and act-tokens, 45–7
aesthetic ideas, *see under* Kant
Ames, William, 175
animal intelligence, 1–2
Anscombe, G.E.M., *Intention*, **73–4**, 81, 84, **101–5**, 119–20, 122–3, 184
Arabian Nights, The, 193
Aristotle, 66; *Physics*, 126
associationist psychology, 203
Audi, Robert, 25
augmentation generation, 47
Augustine, St, 206, 208
Austin, J.L., *How to Do Things with Words*, **36–42**

Barth, John, *Lost in the Funhouse*, 206
Barthes, Roland, 19
Baudelaire, Charles, 'L'Homme et la mer', 108; 'Le Crépuscule du soir', 108
Beardsley, Monroe C., *see under* intentional fallacy
beauty, free and dependent (in Kant), 160–1, 169
Beckett, Samuel, *The Unnamable*, 206
Beddoes, Thomas, 129
Beer, Gillian, *Darwin's Plots*, 172
Belleforest, François de, *Histoires Tragiques*...(source for *Hamlet*), 111
Bergman, Ingmar, *Fanny and Alexander*, **231–4**
Blake, William, 'The Fly', 50
Borges, Jorge Luis, 'Pierre Menard, Author of the *Quixote*', 231
bow-and-arrow metaphor (of purpose), 166
Bridges, Robert, 'Poor Poll', *The Testament of Beauty*, 225
Browning, Robert, 217–18, 224; 'My Last Duchess', 218
Burke, Sean (on deconstruction), 19
Butor, Michel, 13
Byron, George Gordon, Lord, *Don Juan*, 94

Calvino, Italo, *If on a winter's night a traveller*, 9; *Cosmicomics*, 20
Camus, Albert, *The Outsider*, 27–9
causal generation, 47
Cervantes, Miguel de, *Don Quixote*, 230–1
Cinthio, Giovanni Battista Giraldi, *Hecatommithi* (source for *Othello*), 76, 113
Clark, Kenneth (on Leonardo da Vinci), 222
'Co-construction', 235n
Codice Atlantico, 220–2
Coleridge, Samuel Taylor, 41–2, 80, 120, 154 ('Dejection: an Ode'; 'Kubla Khan'), 230; on Kant, **128–37**, 153–4, 162; *Aids to Reflection*, 129; *Biographia Literaria*, 128, 129, 132, 134–5; *Essay on Taste*, 130; *The Friend*, 129; *On the Principles of Genial Criticism*, 131, 136
Collingwood, R.G., 151
Compton-Burnett, Ivy, 170
constative utterances, 36
conventional force (of illocutionary acts), 37–9, 41
conventional generation, 47
Copleston, Frederick (Copleston's rose), 149, 229
Crabb Robinson, Henry, 130
Crawford, D.W. (on Kant), 152–3, 155
Croce, Benedetto, 161
Crow, Christine (on Valéry), 216
Cunningham, J.V., 4

Davie, Donald (on Prince), 213–15
deconstruction, 13–20,
Defoe, Daniel, 192
deliberation (distinct from intention), 63
Derrida, Jacques, 13–17, 19, 235–6 n
Descartes, René, 30–1
Dickens, Charles, 192; *Oliver Twist*, 164
Ding-an-sich, see under *noumenon*
Donne, John, 'The Canonization', 50
Dowling, William C., *see under* New Pragmatics
dramatic monologue, 217–20
driving test (Goldman), 46–8

Index

Eichenbaum, Boris, 90
Eliot, George, *Middlemarch*, 115–18
Eliot, T.S., 17–18, 134, 164, 166, 219–20, 223
elliptical (and non-elliptical) forms of expression, 81
Endzweck, 127

Fichte, Johann Gottlieb, 134
Firbank, Ronald, 170
Fish, Stanley (on *Paradise Lost*), 175
Flaubert, *Letters*, 185
Form of finality (in Kant), 138, 142, 147–52, 154
Fort! Da! (Freud's), 14
Foucault, Michel, 19
Freud, Sigmund, 13–14, 236n

Ginet, Carl, 240n
Golding, William, *Darkness Visible*, 166; *The Spire*, 150
Goldman, Alvin I., *A Theory of Human Action*, **44–50**, 119–20
Goodman, Nelson, 148
Graves, Robert, 'The Florist Rose', 151 'To Evoke Posterity', 187
Green, Henry, 170
Green, J.H. (letter from Coleridge), 137
Grice, Paul, 36, 39, 68, 158, 238–9 n
Griffiths, Eric (on Victorian poetry), 23
Gunn, Thom, **5–10**, 13, 18, 51–2; 'Human Condition', 5–6; 'Vox Humana', 6; 'On the Move', 6; 'Touch', 7–9, 18; 'Back to Life', 9–10, 18
Gustafson, Donald, 25, 241n, 242n

Haller, Albrecht von, 161
Hampshire, Stuart, *Thought and Action*, 2, 20, 109, 246 n
Hegel, Georg Wilhelm Friedrich, *The Philosophy of Fine Art*, 135, 244n
Heidegger, Martin, *Being and Time*, 128, 243 n
Heywood, Jasper (tr. of Seneca), 93
Hill, Geoffrey, 17,
Holloway, John, 11–12, 13, 52, 163
Hooker, Thomas, 175
Hopkins, Gerard Manley, 149–50
hors-texte, 14
Housman, A.E., 'The Culprit', 189
Hughes, Richard, *The Human Predicament*, 211
Hughes, Ted, *Orghast*, 4
Husserl, Edmund, 19–20

illocutionary acts, 37–43, 228
infinitive (of the verb 'to be'), 107–9; (purposive, in Milton), 176–9, 183–4, 186–7
intention, three levels of (in *Macbeth*), 57
intentional fallacy, **21–9**, 66, 81, 120
intentional motive, 117–19, 121–2, 190
intentional/unintentional actions, 102–6
intentions in doing/intentions of doing, 122, 127, 167–8
intentions, multiple and single, 72–3
Ishiguro, Kazuo, *The Unconsoled*, 231

James, Henry, 115–17, 122
Johnson, B.S., *Albert Angelo*, 206
Johnson, Samuel 91, 104 (on *Hamlet*)
Joyce, James, *A Portrait of the Artist as a Young Man*, 185

Kafka, Franz, 'The Truth about Sancho Panza', 230–1; *The Castle*, 231
Kant, Immanuel, 3–4, 12, 235n *Critique of Judgement*, xii, 4, 127, **137–63**: aesthetic ideas, 4, 152–4; structure of the book, 137–9; faculties of understanding and reason, 139–41; faculty of judgement, 141–3; the four moments of taste, 143, 147; judgements of taste, 143–4; 'presentations', 144–5; the imagination, 145–6; the antinomy of taste, 146–7; the third moment of taste ('the form of the purposiveness of an object'), 147–52, 154–7; nature and art, 148–51; free and dependent beauty, 162–3 *Critique of Practical Reason*, 129, 134, 138–40, 153 *Critique of Pure Reason*, 129, 132, 133–4, 138–9, 145, 153; *see also under* Coleridge (on Kant)
Keats, John, 'Ode to a Nightingale', 216
Kenny, Anthony, 44–6, 102
Knapp, Steven, *see under* New Pragmatics
Körner, Stephan (on Kant),
Kyd, Thomas (lost *Hamlet*), 111

La Fontaine, Jean de, 'The Fox and the Goat', 212
Labov, William, on 'natural' and 'literary' narrative, 239n
Lacan, Jacques, 13–16, 30, 134
Leonardo da Vinci, 220–4
level-determinate acts, 50
level-generation, 46–8
Levin, Richard (on *Hamlet*), 95

Lewis, C.S., *A Preface to Paradise Lost*, 230
Livingstone Lowes, J. (on Coleridge), 22, 120
locutionary acts, 37

MacIntyre, Alasdair, 33
MacLeish, Archibald, 'Ars Poetica', 42
Maeterlinck, Maurice, *Pelléas et Mélisande*, 4
Marquez, Gabriel Garcia, *Chronicle of a Death Foretold*, 211
'meaning to', 65
Mellard, James (on Lacan), 14
McCloskey, Mary (on Kant), 147, 163
Michaels, Walter Ben, *see under* New Pragmatics
Milton, John, *De Doctrina Christiana*, 175; *Paradise Lost*, 54, 153, **172–89**, 207–9, 212
Montefeltro, Federico, 249n
Moritz, Karl Philipp, 137
motive, 155ff; in relation to intention, 81–2; *see also* 'intentional motive' and 'psychological motive'
motive-adverbs, 33
motive and action, 239n
'motive' to 'action' sequence, 66, 73
Mozart, Wolfgang Amadeus, *The Marriage of Figaro*, 52
Muir, Kenneth (ed. of *Macbeth*), 54–5
musique concrète, 161

Nagel, Thomas, 235n
New Pragmatics, 25–7
noumenon (*Ding-an-sich*/thing-in-itself), 132, 133–5, 139
nouveau roman, 13, 18–19, 206
Nuttall, A. D., 27–9

Ohmann, Richard, on illocutionary force, 238n

Paley, William, *Natural Theology*, 155–6
parasitical speech acts, 39
paying the builder, 122–5
Payne, Michael (on Lacan), 15–17
performance verbs, *see under* Kenny, Anthony
performative utterances, 36
perlocutionary acts, 37, 41, 228
Poe, Edgar Allan, 'The Raven', 203
poésie pure, la, 161
poisoning the fascists (Anscombe), 73–4
Poole, Thomas (letter from Coleridge), 128

Pope, Alexander, 161
Pound, Ezra, 217, 223
Prince, F. T., 'An Epistle to a Patron', **213–29**
Princip, Gavrilo, 85–7
prior intention, 167–8, 210
psychological motive, 118–20, 190

Rabelais, François, *Gargantua and Pantagruel*, 193
reader-response criticism (Hirsch, Juhl, Fish), 25
're-animation', 11–12, 52
Resnais, Alain, *L'Année dernière à Marienbad*, 19
Richards, I. A., 30, 41–2
Ricks, Christopher (on Clarendon), 20
Robbe-Grillet, Alain, *see under nouveau roman*
Robert, Marthe, *The Old and the New*, 231
Rodway, Allan, 23
Rorty, Richard, 236–7 n
Ryland, John (letter from Coleridge), 129
Ryle, Gilbert, *The Problem of Mind*, **29–36**

Said, Edward, 20
Saussure, Ferdinand de, 16
sawing a plank (Anscombe), 102–4
Saxo Grammaticus (source for *Hamlet*), 111
Schelling, Friedrich Wilhelm, 135
Schiller, Johann Christoph Friedrich von, *Wallenstein*, 91
Schumann, Robert Alexander, 161
Scott, Paul, *The Raj Quartet*, 211
Scruton, Roger (on Kant and Milton), 153; (on aesthetic ideas), 162–3
Searle, John, 36–7; *Intentionality*, 85–7, 102–4, 167–8, 191
Seneca, *Troas*, 93
Sewell, Elizabeth, *The Structure of Poetry*, 214
Sforza, Ludovico, 220, 227
Shakespeare, William,
All's Well that Ends Well, 111
Antony and Cleopatra, 98
Julius Caesar, 100
Hamlet, 66, 90–1, **91–101** (Act 1), **106–10** ('To be or not to be'), 111–14, 117, 118, 165, 168–9, 222–3 (in Bergman's *Fanny and Alexander*)
Henry V, 100

Index

Macbeth, **54–66, 83–91**, 168–9
Measure for Measure, 54
Othello, **67–82**, 113–14, 122, 123–6, 165, 168–9
Troilus and Cressida, 172
Shelley, Percy Bysshe, *Peter Bell the Third*, 225
Simon, Claude, 18
Sophocles, *Oedipus the King*, 103
speech acts, *see under* Austin, J. L.
Spivak, Gayatri Chakravorty, 13
static verbs, *see under* Kenny, Anthony
Sterne, Laurence, *The Life and Opinions of Tristram Shandy*, **189–206**, 210
Strauss, Richard, 161
Strawson, P. F., 1, 39–40, 42, 43
structuralism, 13
subject-identity, 13–14
subjects ('accusative' and 'nominative'), 216–17
suprematism, 161

Tallis, Raymond, 14, 19–20, 104
Tennyson, Alfred, Lord, 'Ulysses', 187
Thelwall, John (letter from Coleridge), 128
thing-in-itself, see under *noumenon*
Tillyard, E. M. W. (on *Paradise Lost*), 175
Tolstoy, Leo, *War and Peace*, 164, 211
tone poems, 161
'truth-gaps', *see* Gustafson

unchanging human heart, doctrine of the, 230–1
Unsworth, Barry, *Morality Play*, 62

Valéry, Paul, 4, 214–16, 226–7
Vygotsky, L. S., 90–1, 97, 157–8

Walcott, Derek, *Omeros*, 230
Warnock, Mary, 153
Wedgwood, Josiah (Coleridge's letters to), 128
Wellek, René, 130
Westminster Assembly, 175
'what for?' / 'why?', 118ff
Wheldon, David, *The Viaduct*, 231
Williams, Nigel, *The Wimbledon Poisoner*, 23
Williams, William Carlos, 42
Wilson, John Dover (on *Hamlet*), 93
Wimsatt, R. K., *see under* intentional fallacy
Wittgenstein, Ludwig, xii, 12–13, 24, 25, 52, 63–5, 134, 150, 235 n
Wollheim, Richard, 145, 148, 237 n.
Wordsworth, William, *The Excursion*, 135–6

Yeats, W. B., 'Among School Children', 150–1

Zweck(mässigkeit), 127, 137